MATTER
Building Block of the Universe

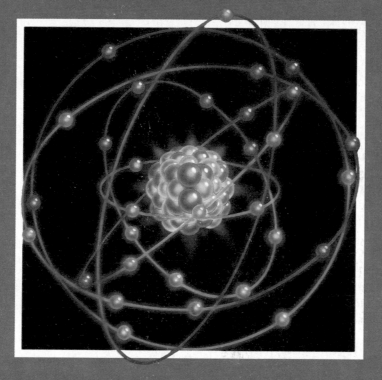

Anthea Maton
Former NSTA National Coordinator
Project Scope, Sequence, Coordination
Washington, DC

Jean Hopkins
Science Instructor and Department Chairperson
John H. Wood Middle School
San Antonio, Texas

Susan Johnson
Professor of Biology
Ball State University
Muncie, Indiana

David LaHart
Senior Instructor
Florida Solar Energy Center
Cape Canaveral, Florida

Maryanna Quon Warner
Science Instructor
Del Dios Middle School
Escondido, California

Jill D. Wright
Professor of Science Education
Director of International Field Programs
University of Pittsburgh
Pittsburgh, Pennsylvania

Prentice Hall
Englewood Cliffs, New Jersey
Needham, Massachusetts

Prentice Hall Science

Matter: Building Block of the Universe

Student Text and Annotated Teacher's Edition
Laboratory Manual
Teacher's Resource Package
Teacher's Desk Reference
Computer Test Bank
Teaching Transparencies
Product Testing Activities
Computer Courseware
Video and Interactive Video

The illustration on the cover, rendered by Keith Kasnot, shows the nucleus of an atom with its protons and neutrons, surrounded by electrons in the electron cloud.

Credits begin on page 172.

SECOND EDITION

ISBN 0-13-402082-0

3 4 5 6 7 8 9 10   97 96 95 94

Prentice Hall
A Division of Simon & Schuster
Englewood Cliffs, New Jersey 07632

STAFF CREDITS

Editorial:	Harry Bakalian, Pamela E. Hirschfeld, Maureen Grassi, Robert P. Letendre, Elisa Mui Eiger, Lorraine Smith-Phelan, Christine A. Caputo
Design:	AnnMarie Roselli, Carmela Pereira, Susan Walrath, Leslie Osher, Art Soares
Production:	Suse F. Bell, Joan McCulley, Elizabeth Torjussen, Christina Burghard
Photo Research:	Libby Forsyth, Emily Rose, Martha Conway
Publishing Technology:	Andrew Grey Bommarito, Deborah Jones, Monduane Harris, Michael Colucci, Gregory Myers, Cleasta Wilburn
Marketing:	Andrew Socha, Victoria Willows
Pre-Press Production:	Laura Sanderson, Kathryn Dix, Denise Herckenrath
Manufacturing:	Rhett Conklin, Gertrude Szyferblatt

Consultants

Kathy French	National Science Consultant
Jeannie Dennard	National Science Consultant
Brenda Underwood	National Science Consultant
Janelle Conarton	National Science Consultant

CONTENTS

MATTER: BUILDING BLOCK OF THE UNIVERSE

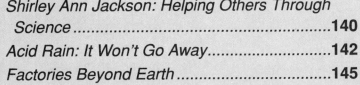

SCIENCE GAZETTE

Activity Bank/Reference Section

Features

CONCEPT MAPPING

Throughout your study of science, you will learn a variety of terms, facts, figures, and concepts. Each new topic you encounter will provide its own collection of words and ideas—which, at times, you may think seem endless. But each of the ideas within a particular topic is related in some way to the others. No concept in science is isolated. Thus it will help you to understand the topic if you see the whole picture; that is, the interconnectedness of all the individual terms and ideas. This is a much more effective and satisfying way of learning than memorizing separate facts.

Actually, this should be a rather familiar process for you. Although you may not think about it in this way, you analyze many of the elements in your daily life by looking for relationships or connections. For example, when you look at a collection of flowers, you may divide them into groups: roses, carnations, and daisies. You may then associate colors with these flowers: red, pink, and white. The general topic is flowers. The subtopic is types of flowers. And the colors are specific terms that describe flowers. A topic makes more sense and is more easily understood if you understand how it is broken down into individual ideas and how these ideas are related to one another and to the entire topic.

It is often helpful to organize information visually so that you can see how it all fits together. One technique for describing related ideas is called a **concept map**. In a concept map, an idea is represented by a word or phrase enclosed in a box. There are several ideas in any concept map. A connection between two ideas is made with a line. A word or two that describes the connection is written on or near the line. The general topic is located at the top of the map. That topic is then broken down into subtopics, or more specific ideas, by branching lines. The most specific topics are located at the bottom of the map.

To construct a concept map, first identify the important ideas or key terms in the chapter or section. Do not try to include too much information. Use your judgment as to what is really important. Write the general topic at the top of your map. Let's use an example to help illustrate this process. Suppose you decide that the key terms in a section you are reading are School, Living Things, Language Arts, Subtraction, Grammar, Mathematics, Experiments, Papers, Science, Addition, Novels. The general topic is School. Write and enclose this word in a box at the top of your map.

SCHOOL

Now choose the subtopics—Language Arts, Science, Mathematics. Figure out how they are related to the topic. Add these words to your map. Continue this procedure until you have included all the important ideas and terms. Then use lines to make the appropriate connections between ideas and terms. Don't forget to write a word or two on or near the connecting line to describe the nature of the connection.

Do not be concerned if you have to redraw your map (perhaps several times!) before you show all the important connections clearly. If, for example, you write papers for Science as well as for Language Arts, you may want to place these two subjects next to each other so that the lines do not overlap.

One more thing you should know about concept mapping: Concepts can be correctly mapped in many different ways. In fact, it is unlikely that any two people will draw identical concept maps for a complex topic. Thus there is no one correct concept map for any topic! Even

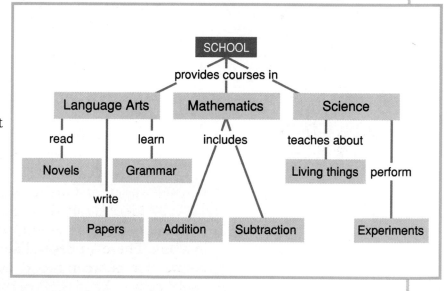

though your concept map may not match those of your classmates, it will be correct as long as it shows the most important concepts and the clear relationships among them. Your concept map will also be correct if it has meaning to you and if it helps you understand the material you are reading. A concept map should be so clear that if some of the terms are erased, the missing terms could easily be filled in by following the logic of the concept map.

MATTER

Building Block of the Universe

Five hundred years ago, Christopher Columbus began a perilous journey. This journey, financed by Queen Isabella of Spain, was designed to find a new route to the riches of the Far East. In those lands, valuable spices and other treasures could be found. The rest is history: In traveling west to find the East, Columbus bumped into a New World.

People have always explored the unknown and made journeys to distant places—if only in their dreams. But no matter how far the journey or how strange the place, one observation can always be made: The universe—as

◀ *In 1492, Christopher Columbus and a brave crew crossed the Atlantic Ocean on a voyage of exploration. This is a woodcut of one of the ships.*

Today's voyagers leave the comforts of city and country to begin to explore the vastness of space. ▶

CHAPTERS

near as one's feet and as far as the most distant star in a galaxy far, far away—is made of the same (and relatively few) kinds of materials.

In this textbook you will begin your own kind of voyage. Your goal is to discover the nature of materials that form you and your universe—the matter inside and around you.

Matter surrounds everyone—whether you share space with other people in a large city or live in a house on a quiet country road.

Discovery *Activity*

Is Something the Matter?

1. Make a list of all the things that are found in an aquarium. Describe each item on your list. If you like, you can set up a small aquarium in your classroom. If you do, however, keep in mind that you are responsible for caring for the animals and plants that live there.

2. As you learn more about the matter that makes up the world around you, add to your list of things that are found in your aquarium.

3. Compare your "before" and "after" lists.
   ■ Has matter taken on any new meanings for you?

General Properties of Matter

On July 19, 1545, a fleet of British warships sailed slowly out of Portsmouth Harbor, England, on its way to battle the French fleet. One ship, the *Mary Rose*, carried a crew of 415 sailors, 285 soldiers, and a number of very new, and very heavy, bronze cannons.

But the *Mary Rose* never met the French fleet. As the story goes, a gust of wind tipped over the *Mary Rose,* and in seconds the ship sank to the bottom of the sea. Was this the true story?

In 1965, teams of scuba divers began a search for the wreck of the *Mary Rose.* Some of the divers wore heavy weights on their belts so that they could hover above the sandy ocean bottom.

In 1970, the *Mary Rose* was found. And scientists uncovered the cause of the ship's sinking. The weight of the heavy bronze cannons had made the ship top-heavy. When the ship tipped over, water rushed into its open spaces, replacing the air. Without the air inside it, the *Mary Rose* had sunk like a stone.

But another mystery remains in this story—a mystery for you to solve. How can a diver wearing a weighted belt hover in the sea, while a ship weighted with cannons and excess water sinks to the bottom? You will uncover the solution as you read on.

Journal *Activity*

You and Your World   Do you know how to swim? Think back to your first few attempts at floating in a pool or lake. What were your feelings as you moved through the water? Write your feelings in your journal. Have you ever swum in the ocean? Is it easier to remain afloat in salt water?

◄ *The secrets of the* Mary Rose—*hidden for so long under water—are revealed by a flashlight's piercing beam.*

1–1 Matter

Taking a bit of the Earth's air along, the astronaut you see in Figure 1–1 is walking over the surface of the moon. He and his fellow astronauts traveled a great distance on their journey to the moon and back, and, fortunately for us, they did not return home empty-handed. For along with tales of triumph, they brought back some of the moon itself: moon rocks for scientists to study in a laboratory and a special piece of moon rock for all to touch. This special rock, once part of the moon's surface, is now one of the great treasures on display at the Smithsonian Institution in Washington, DC. Touch this rock and your mind can journey to the moon with brave astronauts. Touch this rock and you can feel the stuff of the universe. But did you know that you can touch the stuff of the universe right here on Earth?

What Is Matter?

You see and touch hundreds of things every day. And although most of these things differ from one another, they all share one important quality: They are all forms of **matter.** Matter is what the universe is made of. Matter is what you are made of.

Through your senses of smell, sight, taste, and touch, you are familiar with matter. Some kinds of matter are easily recognized. Wood, water, salt, clay, glass, gold, plants, animals—even a piece of the moon—are examples of matter that are easily observed. Oxygen, carbon dioxide, ammonia, and air are kinds of matter that may not be as easily recognized. Are these different kinds of matter similar in some ways? Is salt anything like ammonia? Do water and glass have anything in common?

In order to answer these questions, you must know something about the **properties,** or characteristics, of matter. Properties describe an object. Color, odor, size, shape, texture, and hardness are properties of matter. These are specific properties of matter, however. Specific properties make it easy to tell one kind of matter from another. For example, it is

Figure 1–1 *The moon is the first body in space to feel the footprints of a human being. In this photograph you can see the ultimate "dune buggy," a specially designed vehicle that is able to scoot along the moon's soft surface. What kinds of technology make a moon visit possible?*

not hard to tell a red apple from a green one, or a smooth rock from a rough one.

Some properties of matter are more general. Instead of describing the differences among forms of matter, general properties describe how all matter is the same. **All matter has the general properties of mass, weight, volume, and density.**

ACTIVITY WRITING

Describing Properties

In a novel, the author describes the properties of the objects he or she is writing about. These details add interest to the story.

Collect at least six different kinds of objects. You might include rocks, pieces of wood or metal, and objects made by people. Identify each sample by its general properties and by its special properties. Now write a short paragraph that uses the descriptions you have developed. Be sure to include the following properties in your paragraph: color, density, hardness, mass, texture, shape, volume, and weight. Here is an example of the beginning of a paragraph:

It was a cold, wintry night as Jeff walked home from school. Small six-sided snowflakes fell to the ground. Walking past the Jefferson house, a three-story mansion with many pointed window frames, each of which had at least one broken pane of glass, Jeff was startled to see a huge shape. He had heard this house was haunted. . . .

Figure 1–2 *Rocks carved by winds, cascading water, beautiful plants, and floating magnets are all made of matter. In fact, everything on Earth—and beyond—is made of matter.*

1. What is matter?
2. Name four general properties of matter.

Connection—*Astronomy*
3. Imagine that you have voyaged to deep space, far beyond our known universe. There you have encountered a planet inhabited by people who are much like you and who understand your language. They are very curious about Planet Earth. Describe for them the matter that makes up your home.

Guide for Reading

Focus on this question as you read.

▶ *What is the difference between mass and weight?*

1–2 Mass and Weight

The most important general property of matter is that it has **mass. Mass is the amount of matter in an object.** The mass of an object is constant. It does not change unless matter is added to or removed from the object.

Mass, then, does not change when you move an object from one location to another. A car has the same mass in Los Angeles as it has in New York. You have the same mass on top of a mountain as you do at the bottom of a deep mine. In fact, you would have that exact same mass if you walked on the surface

Figure 1–3 *You know that a bowling bowl has weight, for it is heavy to lift. The weight of a bowling bowl depends upon gravity. The weight can change if the force of gravity acting on it changes. Does the mass of a bowling ball ever change?*

Figure 1–4 *It is not really magic—just a demonstration of inertia. As the table on which this dinner is set (left) is moved quickly, the objects are suspended in midair, but only for an instant (right).*

of the moon! Later in this chapter you will discover for yourself why this is such an important concept.

Mass and Inertia

Scientists have another definition for mass. Mass is a measure of the **inertia** (ihn-ER-shuh) of an object. Inertia is the resistance of an object to changes in its motion. Objects that have mass resist changes in their motion. Thus objects that have mass have inertia. For example, if an object is at rest, a force must be used to make it move. If you move it, you notice that it resists your push or pull. If an object is moving, a force must be used to slow it down or stop it. If you try to stop a moving object, it will resist this effort.

Suppose you were given the choice of pushing either an empty shopping cart or a cart full of groceries up a steep hill. The full cart, of course, has more mass than the empty one. And as you might know from past experience, it is much easier to push something that is empty than it is to push something that is full. Now suppose the empty cart and the full

Figure 1–5 *Once a bobsled is moving on the slippery sheet of ice that makes up its run, it can reach high speeds. However, to overcome its inertia and to get the bobsled moving requires the strength of four strong people.*

ACTIVITY

DISCOVERING

Demonstrating Inertia

You can demonstrate that objects at rest tend to remain at rest by using a drinking glass, an index card, and a coin.

1. Place the glass on a table.

2. Lay a flat index card on top of the glass. Place the coin in the center of the card.

3. Using either a flicking motion or a pulling motion of your fingers, quickly remove the card so it flies out from under the coin. Can you remove the card fast enough so the coin lands in the glass? You might need to practice a few times.

How does this activity demonstrate inertia?

■ What happens to the coin if you remove the card slowly?

How does removing the card slowly demonstrate inertia?

cart are at the top of the hill and begin to roll down. Again, you might know from experience that the full cart—the cart with more mass—will be more difficult to stop than the empty cart. In other words, it is more difficult to get the cart with more mass moving and it is more difficult to get it to stop.

The more mass an object has, the greater its inertia. So the force that must be used to overcome its inertia also has to be greater. That is why you must push or pull harder to speed up or slow down a loaded shopping cart than an empty one.

Mass is measured in units called grams (g) and kilograms (kg). One kilogram is equal to 1000 grams. The mass of small objects is usually measured in grams. The mass of large objects is usually measured in kilograms. For example, a nickel has a mass of about 5 grams. The mass of an average-sized textbook is about 1600 grams, or 1.6 kilograms. The mass of an elephant may be more than 3600 kilograms.

Weight: A Changeable Property of Matter

In addition to giving an object inertia, mass is also the reason an object has **weight.** Weight is another general property of matter. If a scientist is asked how much she weighs, she is correct in answering that it depends. This is because weight is not constant. Weight changes according to certain conditions. You probably know that your weight changes. It increases after you eat a large meal. It decreases after you spend time exercising. In these cases, your mass also changes. It increases when you eat and decreases when you exercise and burn off Calories. (Remember, mass can only change if matter is added to an object or taken away from an object.) But weight can change even when an object's mass remains the same. An object's weight, unlike its mass,

Figure 1–6 *The mass of a harvest mouse balancing itself on some strands of wheat is so small that it is measured in grams. The polar bear is a different story. Its huge mass is measured in kilograms—and many of them. Do you know the metric unit used to measure weight?*

Figure 1–7 *It is the force of Earth's gravity that determines exactly what you weigh. The same person would weigh less on a mountaintop than in a mine deep within the Earth. Why would a person weigh more in a mine than on a mountaintop?*

Figure 1–8 *It seems such a simple act: keeping three balls suspended in space at the same time. Remember, though, that the juggler is always acting against the force of gravity—which would surely cause the balls to fall to the ground at the first mistake.*

is also dependent on its location (where it is). In order to understand what weight is and why it is not constant, you must know something about gravity.

Weight and Gravity

You have probably noticed that a ball tossed up in the air falls to the ground. This happens regardless of how hard the ball is thrown. You also know that an apple that drops from a tree falls down to the ground, not up in the air. Both the ball and the apple fall down because of the Earth's force of attraction for all objects. The force of attraction between objects is called **gravity.**

Gravitational force is not a property of the Earth alone. All objects exert a gravitational attraction on other objects. Indeed, your two hands attract each other, and you are attracted to books, papers, and chairs. Why then are you not pulled toward these objects as you are pulled toward the Earth? In fact, you are! But the attractions between you and the objects are too weak for you to notice them. The Earth's gravity, however, is great because the Earth is so massive. In fact, the greater the mass of an object, the greater its gravitational force. How do you think the gravity of Jupiter, many times more massive than Earth, compares with that of Earth?

The pull of gravity on an object determines the object's weight. On the Earth, your weight is a direct measure of the planet's force pulling you toward the center. The pull of gravity between objects weakens as the distance between the centers of the objects becomes greater. So at a high altitude—on top of a tall mountain, for example—you actually weigh less than you do at sea level. This is because you are farther from the center of the Earth on top of a high mountain than you are at sea level. Remember this idea if you are ever on a diet. Bring a bathroom scale to the top of Mount Everest, the highest mountain on Earth, and weigh yourself there. There is no place on Earth where you will weigh less.

When an object is sent into space far from Earth, the object is said to become weightless. This is because the gravitational force of the Earth on the object decreases as the object moves away from the

A Quick Weight Change

An inhabitant of Planet X weighs 243 eigers on her home planet. The gravity of Planet X is 2.7 times greater than that of Earth. How many eigers will she weigh on Earth?

Figure 1–9 *In the nineteenth century, space travel was only a dream in a writer's mind. The illustration is from a book written by the great French writer Jules Verne. Today, however, space travel is real—as this astronaut floating untethered high above the Earth's surface demonstrates. Even though this astronaut appears weightless, has his mass changed?*

Scuba Diver

At the beginning of this chapter you read about the divers who searched for the *Mary Rose.* Those divers were specially trained **scuba divers** who earn their living by completing underwater tasks with the use of scuba gear.

The tasks of scuba divers differ from job to job. Their work may include inspecting dams, pipelines, or cables. Or it may involve recovering valuable sunken objects or doing underwater repair and construction work.

If you are interested in scuba diving as a career, you can obtain more information by writing to PADI, 1251 East Dyer Road #100, Santa Ana, CA 92705.

Guide for Reading

Focus on this question as you read.

▶ *How can you determine the density of an object?*

center of the Earth. An object in space is a great distance from the center of the Earth—in some instances, millions of kilometers. However, although the object is said to be weightless, it really is not. The gravitational force of the Earth decreases, but it still exists.

Although an object in space is said to become weightless, it *does not* become massless. Mass, remember, does not change even though location changes. So no matter what happens to the force of gravity and the weight of an object, its mass stays the same. Only its weight can change.

The metric unit of weight is the newton (N). The newton is used because it is a unit of force, and weight is the amount of force the Earth's gravity exerts on an object. An object with a mass of 1 kilogram is pulled to the Earth with a force of 9.8 newtons. So the weight of this object is 9.8 N. An object with a mass of 50 kilograms is pulled toward the Earth with a force of 50 times 9.8, or 490 newtons. The object's weight is 490 N.

1–2 Section Review

1. What is mass? What is weight?
2. How are mass and inertia related?

Critical Thinking—*Applying Concepts*
3. The moon is smaller than the Earth. Where would you weigh less? Where would you have less matter?

1–3 Volume and Density

Let's use this textbook to help discover another general property of matter. Suppose you could wrap a piece of paper around this entire book and then remove the book inside. How would you describe what was left inside the paper? You would probably use the word space. For an important property of matter is that it takes up space. And when the book

CUBIC CENTIMETER
(cc or cm³)

1 cm

1 cm   1 cm

is occupying its space, nothing else can be in that same space. You might prove this to yourself.

The amount of space an object takes up is called its **volume.** The metric units that are used to express volume are the liter (L), milliliter (mL), and cubic centimeter (cm³). In general, liters and milliliters are used to measure the volume of liquids, and cubic centimeters are used to measure the volume of solids. One milliliter is equal in volume to one cubic centimeter. One thousand milliliters is equal to one liter. How many milliliters are there in 2.5 liters?

Volume is an important property of matter that you use every day. Many products you may buy at a store (milk and bottled water, for example) are sold in liter containers. Cough syrups and many prescription drugs are measured in milliliters. Although you may not see cubic centimeters as frequently, you would certainly need this unit of measurement to describe the volume of a set of wooden blocks you might want to purchase for a younger brother or sister.

You now know two important general properties of matter: Matter has mass and it occupies space. You can use these two properties to define matter in a more scientific way: **Matter is anything that has mass and volume.**

The properties of mass and volume can be used to describe another important general property of matter called **density.** Density is often used to describe things. A pine forest is often called a dense forest if the trees grow close together. You may have said your best friend was dense when he or she did not understand a joke you told. To a scientist, density has a specific meaning. **Density is the mass per unit volume of an object.**

Density is an important property because it allows you to compare different types of matter. Let's see how. Suppose you were asked to determine which is heavier, wood or steel. How would you go about doing it? Perhaps you would suggest comparing the masses of both on a balance. You are on the right track, but there is one problem with this solution. What size

ACTIVITY
DISCOVERING

Volume of a Solid

You can easily measure the volume of a liquid by using a graduated cylinder. Can this method be used to determine the volume of a solid?

Fill a graduated cylinder half full with water. Note the volume of the water. Now place a small solid object in the graduated cylinder. You might choose a rock, a block of wood, or a bar of soap. If the object floats, use a piece of wire to push it under the water's surface. Note the new volume of the water.

You now have two volumes for the liquid—the original volume and the new volume. Ask yourself these questions to find the volume of the solid:

What caused the change in volume?

Is the volume change different for different objects?

■ How is the change in liquid volume related to the volume of the solid object?

DENSITIES OF SOME COMMON SUBSTANCES	
Substance	**Density (g/cm³)**
Air	0.0013
Gasoline	0.7
Wood (oak)	0.85
Water (ice)	0.92
Water (liquid)	1.0
Aluminum	2.7
Steel	7.8
Silver	10.5
Lead	11.3
Mercury	13.5
Gold	19.3

Figure 1–11 *This chart shows the densities of some common substances. Which substances would float on water? Which ones would sink?*

pieces of wood and steel would you use? After all, a small piece of steel might have the same mass as a large piece of wood.

You are probably beginning to realize that in order to compare the masses of two objects, you need to use an equal volume of each. When you do, you soon discover that a piece of steel has a greater mass than a piece of wood *of the same size.* And that is the important part of that statement—of the same size. So for our example we can say a cubic centimeter of steel is heavier than a cubic centimeter of wood. Or steel is denser than wood.

All matter has density. And the density of a specific kind of matter is a property that helps to identify it and distinguish it from other kinds of matter.

Since density is equal to mass per unit volume, we can write a formula for calculating the density of an object:

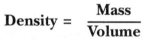

$$\text{Density} = \frac{\text{Mass}}{\text{Volume}}$$

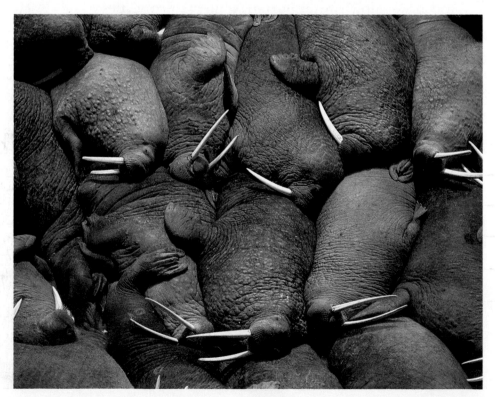

Figure 1–12 *Biologists use the term population density to refer to the number of individual organisms in a given area. The population density of these walruses on a crowded beach would prevent even one more walrus from squeezing in.*

Density is often expressed in grams per milliliter (g/mL) or grams per cubic centimeter (g/cm³). The density of wood is about 0.8 g/cm³. This means that a piece of wood 1 cubic centimeter in volume has a mass of about 0.8 gram. The density of steel is 7.8 g/cm³. So a piece of steel has a mass about 9.75 times that of a piece of wood of the same size.

The density of fresh water is 1 g/mL. Objects with a density less than water float. Objects with a density greater than water sink. Thus wood floats in water because its density is less than the density of water. What happens to a piece of steel when it is put in water?

If you have ever placed an ice cube in a glass of water, you know that ice floats. So frozen water (ice) must be less dense than liquid water. Actually, the density of ice is about 89 percent that of cold water. What this means is that only about 11 percent of a block of ice stays above the surface of the water. The rest is below the surface. This fact is what makes icebergs so dangerous. For it is only the "tip of the iceberg" that is visible.

Activity Bank

What "Eggs-actly" Is Going on Here?, p.150

Sample Problem

If 96.5 grams of gold has a volume of 5 cubic centimeters, what is the density of gold?

Solution

Step 1	Write the formula.	$Density = \dfrac{Mass}{Volume}$
Step 2	Substitute given numbers and units.	$Density = \dfrac{96.5 \text{ grams}}{5 \text{ cubic centimeters}}$
Step 3	Solve for unknown variable.	$Density = \dfrac{19.3 \text{ grams}}{\text{cubic centimeter}}$

Practice Problems

1. If 96.5 g of aluminum has a volume of 35 cm³, what is the density of aluminum? How does its density compare with the density of gold?

2. If the density of a diamond is 3.5 g/cm³, what would be the mass of a diamond whose volume is 0.5 cm³?

ACTIVITY THINKING

You may have read or heard about the passenger ship *Titanic*, which sank in 1912 after it ran into an iceberg in the cold North Atlantic. The most advanced technology was used to build the *Titanic*. Special watertight doors were designed so that they could seal off a part of the ship that developed a leak. The ship was said to be unsinkable. However, its watertight doors were not able to keep the ocean waters from filling the *Titanic* when the iceberg ripped through the side of the ship. Once ocean water replaced the air in the *Titanic*, the density of the ship was no longer less than the density of water and the ship plunged to the ocean bottom on its maiden voyage!

Can you now solve the mystery posed at the beginning of this chapter? An object floats in water if its density is less than 1 g/mL. In order for the

Figure 1–13 *The "unsinkable"* Titanic *sank after striking an iceberg in the North Atlantic. As it filled with water—through the gaping hole in its hull caused by the iceberg—the density of this great ship became greater than the ocean water upon which it floated. And it sank under the waves.*

scuba diver searching for the *Mary Rose* to sink in the water, the diver's overall density has to be greater than 1 g/mL. So the diver wears a weighted belt to increase mass.

The density of water increases as the temperature of the water gets colder. Below the ocean's surface, the temperature of the water decreases. So the density of deep cold water is greater than 1 g/mL. At a certain depth, the scuba diver's density is equal to the water's density. The diver will not be able to sink below this depth.

While the *Mary Rose* moved on the surface of the ocean, her hull was partly filled with air. The air helped make the ship's density less than 1 g/mL, and so it floated. The large volume of air balanced the added mass of the heavy bronze cannons. However, when her hull partially filled with water, the *Mary Rose* and her heavy cannons became denser than the surrounding water at any depth. Down, down went the *Mary Rose!*

Figure 1–14 *Because of the air within it, a huge ocean liner can float on the surface of the ocean (left). By pumping water into and out of special tanks, submarines are able to sink or float at will (right). These fish can maintain their position in the water by emptying or filling an air bladder within their body (inset). How does a life preserver help a swimmer remain afloat?*

Figure 1–15 *Unlike all the other planets in our solar system, magnificent Saturn has a density less than 1 g/mL. In fact, if you could find a large enough ocean, Saturn would float in it.*

1–3 Section Review

1. What is density?
2. What determines whether an object floats or sinks in water?

Critical Thinking—*Making Comparisons*

3. Perhaps you have attended a party where balloons floated. A balloon filled with air does not rise above your head, but a balloon filled with helium gas rises in the air to the end of its string. What does this tell you about the density of helium?

PROBLEM Solving

A Density Disaster

Look closely at the accompanying photograph. It shows the densities of some common substances. As you can see, some objects float in water and others sink.

Now pretend that this photograph represents a small portion of the ocean. Floating on this ocean is a steel oil tanker filled with crude oil. (Remember that because much of its volume is filled with air, a large ship such as a tanker is less dense than water and thus will float.) Suddenly the tanker runs aground on a reef. A huge hole is torn in the ship's hull. Oil gushes out of the ship into the water. This could be a major environmental disaster!

Cause and Effect

Assume that the crude oil has the same density as corn oil.

1. Why does the oil pose a great danger?
2. Is the danger greater to birds and marine mammals than it is to fish and other organisms that live on the ocean bottom?
3. How is the density of oil an advantage in the cleanup?
4. Why would an oil spill be an even greater disaster if the density of oil were the same as that of corn syrup?

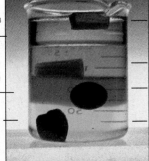

Air — 0.001 g/cm³
Corn oil — 0.93 g/cm³
Water — 1.00 g/cm³
Glycerine — 1.26 g/cm³
Corn Syrup — 1.38g/cm³

Wood — 0.85 g/cm³
Plastic — 1.17 g/cm³
Rubber — 1.34 g/cm³
Steel — 7.81 g/cm³

Up, Up, and Away

For many thousands of years, people have dreamed of flying through the air as gracefully as birds. According to Greek *mythology,* Daedalus and his son, Icarus, escaped from the labyrinth using wings made of feathers, wax, and thread. Icarus, however, flew too close to the sun and the wax melted, plunging this early aviator to his death in the sea.

In the eighteenth century, human flight became a reality, not just mythology. The Montgolfier brothers launched the first hot-air balloon in *history*—and in so doing, captured the imagination of the French people as well. Balloonamania swept France. The principles behind their hot-air ballooning are those you learned about in this chapter.

The balloon developed by the Montgolfiers was made of silk, carefully suspended over a fire. The flames of the fire heated the air in the balloon, causing the air to expand. As the air expanded, it occupied more space and the balloon inflated. In other words, the volume of the air increased. As the volume of the heated air increased, the density of the hot-air balloon became less than the density of the cooler air that surrounded it. The balloon—now lighter than air—rose off the ground and began its historic voyage upward.

Unlike the balloon invented by the Montgolfier brothers, modern hot-air balloons rely on tanks of flammable gas to heat the air. The height of the balloon is controlled by heating the air within the balloon. Of course, even the most avid ballooners eventually want to come back to Earth. To do so, they allow the air in the balloon to cool. When this happens, an opposite reaction occurs. The air contracts and occupies less space. Its volume decreases and it becomes denser than the surrounding air. The balloon, no longer lighter than air, descends. If the pilot wants to descend rapidly, air can also be released from special flaps at the top of the balloon. By controlling the amount of heat as well as the amount of air released, the rate of ascent or descent of the balloon can be carefully regulated.

Today, many people fly hot-air balloons, but it was the pioneering work of the Montgolfier brothers that opened up the world of flight to humans—a world once limited only to birds and beasts.

Brave Icarus tried to fly and failed. However, the hot-air balloon took people to heights undreamed of.

Laboratory Investigation

Inertia

Problem

How does an object's mass affect its inertia?

Materials *(per group)*

several shoe boxes
objects of various masses to fit in the shoe boxes
smooth table top
household broom
meterstick or metric ruler

Procedure

1. Place an object in each shoe box and replace the lid on the shoe box. Number each box. (Your teacher may provide you with several shoe boxes that are already prepared.)
2. Position the box so that it hangs over the edge of the table by 8 cm.
3. Stand the broom directly behind the table. Put your foot on the straw part of the broom to hold it in place.
4. Slowly move the broom handle back away from the box.
5. When you release the handle, the broom-stick should spring forward, striking the middle of the end of the box.
6. Measure how far the box moves across the table after it is struck by the broom.
7. Repeat this procedure with each of the boxes. Try to use the same force each time.

Observations

1. Enter the box number and the distance moved in a chart similar to the one shown here.

Box Number	Distance Traveled

2. Open the boxes and examine the contents. Record what object was in each box.

Analysis and Conclusions

1. What part of the definition of inertia applies to your observations about the movements of the boxes?
2. Why do you think some boxes moved farther than others?
3. What do you notice about the objects that moved farthest from the resting position? What do you notice about the objects that moved the shortest distance from the resting position?
4. Why was it important that you used the same force each time a box was struck?
5. **On Your Own** You can compare the masses of different objects by using a balance. Can you propose another way to determine the masses of different objects?

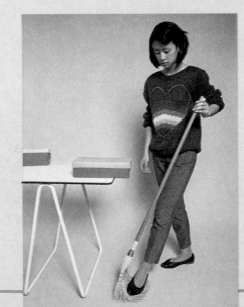

Summarizing Key Concepts

1–1 Matter

▲ All objects are made of matter.

▲ A property is a quality or characteristic that describes matter.

▲ General properties describe how all matter is the same. Specific properties describe the differences among forms of matter.

▲ All matter has the general properties of mass, weight, volume, and density.

1–2 Mass and Weight

▲ One property of matter is that it has mass. Mass is the amount of matter in an object.

▲ The property of matter that resists changes in motion is called inertia. Mass is a measure of the inertia of an object.

▲ Mass is commonly measured in grams or kilograms.

▲ The force of attraction between objects is called gravity.

▲ The gravitational attraction between objects is dependent on their masses.

▲ Gravitational attraction between objects becomes less as the distance between the objects increases.

▲ The pull of gravity on an object determines the object's weight.

▲ The weight of an object can vary with location, but its mass never changes unless matter is added to or taken from the object.

1–3 Volume and Density

▲ The amount of space an object takes up is called its volume.

▲ Volume is measured in liters, milliliters, and cubic centimeters. In general, liters and milliliters are used to measure liquid volumes and cubic centimeters are used to measure solid volumes.

▲ The density of an object is its mass per unit of volume. Density equals mass divided by volume.

▲ The density of a particular kind of matter is a specific property that helps identify it.

▲ The density of liquid water is 1 gram per milliliter (1 g/mL).

▲ Objects that float in water have a density less than 1 gram per milliliter. Objects with a density greater than 1 gram per milliliter sink in water.

Reviewing Key Terms

Define each term in a complete sentence.

1–1 Matter
matter
property

1–2 Mass and Weight
mass
inertia
weight
gravity

1–3 Volume and Density
volume
density

Chapter Review

Content Review

Multiple Choice

Choose the letter of the answer that best completes each statement.

1. Characteristics that describe how all matter is the same are called
   a. specific properties.
   b. universal differences.
   c. density numbers.
   d. general properties.
2. The amount of matter in an object is a measure of its
   a. volume.   c. density
   b. mass.   d. weight.
3. In describing the mass of an object, it is correct to say that
   a. mass changes with altitude.
   b. mass changes with location.
   c. mass remains unchanged.
   d. mass changes with weight.
4. The formula for finding density is
   a. volume/mass.   c. mass/volume.
   b. volume x mass. d. mass/weight.

5. As an object gets farther from Earth,
   a. its weight increases.
   b. its weight decreases.
   c. its mass decreases.
   d. its weight remains the same.
6. The amount of space an object takes up is called its
   a. volume.   c. weight
   b. density.   d. inertia.
7. An object's resistance to a change in motion is called its
   a. density.   c. mass.
   b. inertia.   d. volume.
8. The force of attraction between objects is
   a. inertia.   c. density.
   b. weight.   d. gravity.

True or False

If the statement is true, write "true." If it is false, change the underlined word or words to make the statement true.

1. All objects are made up of <u>matter</u>.
2. <u>Volume</u> is a measure of the resistance of an object to changes in its motion.
3. One liter is equal to <u>100</u> milliliters.
4. Some general properties of matter include <u>mass, weight, color, and volume</u>.
5. <u>Density</u> is the amount of space an object takes up.
6. As an object's weight increases, its mass <u>decreases</u>.
7. An object's mass per unit volume is called its <u>density</u>.
8. An object that floats in water has a density <u>greater</u> than 1 g/mL.

Concept Mapping

Complete the following concept map for Section 1–1. Refer to pages N6–N7 to construct a concept map for the entire chapter.

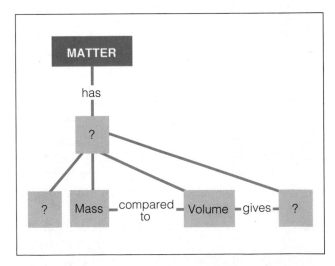

Concept Mastery

Discuss each of the following in a brief paragraph.

1. A rocket taking off from Earth needs much more fuel than the same rocket taking off from the moon. Explain why.
2. Why are astronauts floating above the Earth in a Space Shuttle really not weightless?
3. A person who cannot float in a freshwater lake can float easily in the Great Salt Lake in Utah. What does this tell you about the density of salt water?
4. Each year some college students have a contest to build and race concrete boats. What advice would you give the students to make sure their boats float?
5. Fish are able to remain at a specific depth in water without much trouble. Many fish have an organ called... they can fill with air and e... does a swim bladder help a fish sta... one level in the water?
6. On Earth, a ballet dancer has a great deal of trouble lifting a ballerina over his head. On the moon, however, he can lift her with ease. Explain this situation.
7. An ice cube is only frozen water. Why does an ice cube float on the surface of a glass of water and not sink to the bottom?
8. "Oil and water don't mix" is an old saying. Use what you have learned about density to explain the scientific reasons for this saying.

Critical Thinking and Problem Solving

Use the skills you have developed in this chapter to answer each of the following.

1. **Making comparisons** You are given two samples of pure copper, one with a mass of 20 grams and the other with a mass of 100 grams. Compare the two samples in terms of (a) volume, (b) weight, and (c) density.
2. **Applying concepts** Explain why selling cereal by mass rather than by volume would be more fair to consumers.
3. **Making calculations** If the density of a certain plastic used to make a bracelet is 0.78 g/cm^3, what mass would a bracelet of 4 cm^3 have? Would this bracelet float or sink in water?
4. **Designing an experiment** The common metal iron pyrite (bottom) is often called fool's gold because it can be mistaken for gold (top). Design an experiment to determine whether a particular sample is iron pyrite or real gold.
5. **Making inferences** Aluminum is used to make airplanes. Cast iron is used to make heavy machines. Based on this informa-tion, compare the densities of aluminum and cast iron.
6. **Using the writing process** Write a short poem about matter. Use at least two general properties of matter in your work.

Physical and Chemical Changes

Guide for Reading

After you read the following sections, you will be able to

2–1 Phases of Matter
- Recognize the four phases of matter.
- Describe the gas laws.

2–2 Phase Changes
- Describe the phase changes that occur in matter.
- Relate phase changes to changes in heat energy.

2–3 Chemical Properties and Changes
- Distinguish between a chemical property and a chemical change.

At dawn's first light the weather forecast indicated the day would be sunny and bright. But throughout the day the temperature at the orange grove dropped. By afternoon it was so cold that the workers became concerned about the orange trees whose branches were heavy with fruit not yet ripe enough to be picked. Such low temperatures could wipe out the entire crop.

The workers knew that something had to be done quickly. Some workers lighted small fires in smokepots scattered throughout the groves. But they soon realized that the heat produced in this way would not save the fruit. Suddenly other workers raced into the grove hauling long water hoses! These workers began to spray the trees with water. As the temperature continued to drop, the water would freeze into ice. The ice would keep the oranges warm!

Does it seem strange to you that oranges can be protected from cold by ice? Was this some sort of magic? In a sense, yes. But it was magic that anyone who knows science can do. And as you read further, you will learn how freezing water can sometimes work better than a fire to keep things warm.

Journal *Activity*

You and Your World You know that water is a liquid. You also know that when it is frozen, water can also be a solid. Make a list of the kinds of things that you can do with liquid water. Make a second list of things you can do with solid water. Enter your lists in your journal.

◄ *To keep oranges from being destroyed by freezing temperatures, the oranges are sprayed with water that quickly freezes into ice. How does ice protect the oranges? The answer to this question lies within the pages of this chapter.*

2–1 Phases of Matter

The general properties of matter that you learned about in Chapter 1—mass, weight, volume, and density—are examples of **physical properties.** Color, shape, hardness, and texture are also physical properties. Physical properties are characteristics of a substance that can be observed without changing the identity of the substance. Wood is still wood whether it is carved into a baseball bat or used to build the walls of a house.

Ice, liquid water, and water vapor may seem different to you. Certainly they differ in appearance and use. But ice, liquid water, and water vapor are all made of exactly the same substance in different states. These states are called **phases.** Phase is an important physical property of matter. Scientists use the phases of matter to classify the various kinds of matter in the world. **Matter can exist in four phases: solid, liquid, gas, and plasma.**

Solids

A pencil, a cube of sugar, a metal coin, and an ice cream cone are examples of **solids.** All solids share two important characteristics: Solids have a definite shape and a definite volume. Let's see why. The tiny particles that make up a solid are packed very close together. Because of this arrangement, the particles cannot move far out of their places, nor can they flow over or around one another. In a solid, the tightly packed particles are able only to vibrate. Little other motion occurs. Thus a solid is able to keep its definite shape.

Figure 2–1 *Living up to its name, Old Faithful geyser in Yellowstone National Park erupts on schedule. What phases of water can you observe in this photograph?*

Figure 2–2 *Sodium chloride, or table salt, and potassium chloride, which is sometimes used to season foods by people who must limit the amount of sodium they eat, are two common crystalline solids. The illustration shows how atoms are arranged in a sodium chloride crystal.*

Figure 2–3 *Crystals can vary in color and shape. Valuable ruby crystals are used to make jewelry (left). Gypsum crystals are valuable in their own right (center). Gypsum is used to make wallboard and other construction materials. Fluorite crystals (right), which can be clear or colored, are used as a source of fluorine and in glassmaking.*

If you could examine the internal structure of many solids, you would see that the particles making up the solids are arranged in a regular, repeating pattern called a **crystal.** Solids made up of crystals are called crystalline solids. A good example of a crystalline solid is common table salt. Figure 2–3 shows several other, more colorful examples of crystalline solids.

Crystals often have beautiful shapes that result from the arrangement of the particles within them. Snowflakes are crystals of water in the solid phase. If you look at them closely, you will see that all snowflakes have six sides. However, what is so amazing is that no two snowflakes in the world are ever exactly alike.

There are some solids, however, in which the particles are not arranged in a regular, repeating pattern. These solids do not keep their definite shapes permanently. Because the particles in these solids are not arranged in a rigid way, they can slowly flow around one another. Solids that lose their shape under certain conditions are called amorphous (uh-MOR-fuhs) solids. Have you ever worked with sealing wax or silicone rubber? If so, you have worked with an amorphous solid.

Actually, an amorphous solid can also be thought of as a slow-moving liquid. Candle wax, window glass, and the tar used to repair roads are amorphous solids that behave like slow-moving liquids. You were

Activity Bank

Crystal Gardening, p.152

Figure 2–4 *The computer-generated drawing of a portion of a snowflake shows the repeating pattern of the particles of ice that make up the crystal.*

probably surprised to learn that glass is a slow-moving liquid. Although it moves too slowly to actually observe, you can see the results of its movement under certain conditions. If you look at the window-panes in a very old house, you will notice that they are thicker at the bottom than at the top. Over time, the glass has flowed slowly downward, just like a liquid. In fact, glass is sometimes described as a super-cooled liquid. Glass forms when sand and other materials in the liquid phase are cooled to a rigid condition without the formation of crystals.

Liquids

Although the particles in a **liquid** are close together, they are not held as tightly together as are the particles in a solid. So the particles in a liquid are free to move. Thus a liquid has no definite shape. It takes the shape of its container. A liquid in a square container is square. The same liquid in a round container is round.

Although liquids do not have a definite shape, they do have a definite volume. One liter of water is still one liter of water whether it is in a round container or a square one. And if that one liter of water is poured into a two-liter bottle, it will occupy only half the bottle's volume. It will not fill the bottle. One liter of water does not spread out to fill a two-liter bottle. What do you think would happen if you tried to pour that one liter of water into a half-liter bottle?

Remember that the particles in a liquid are free to move. This movement is basically a flowing around one another. Some liquids flow more easily than others, however. The resistance of a liquid to flow is called viscosity (vihs-KAHS-uh-tee). Honey has a high viscosity compared to water. This means that honey flows more slowly than water. If you have ever poured honey from a jar, you are probably familiar with this fact. The oil you put in an automobile also has a high viscosity. This is important because the oil coats the moving parts in the motor and prevents them from rubbing together and wearing out.

ACTIVITY

DISCOVERING

Observing Viscosity

Remember that some liquids flow more easily than others. Viscosity is the resistance of a liquid to flow.

1. Obtain samples of the following: catsup, corn syrup, milk, honey, maple syrup.

2. Cover a piece of cardboard with aluminum foil.

3. Place the cardboard on a plate or baking pan at about a 45- to 50-degree angle with the bottom of the plate or pan.

4. With four classmates helping you, pour a measured amount of each liquid from the top of the cardboard at the same time.

5. Determine the order in which the liquids reach the bottom of the cardboard.

Which liquid is the most viscous? The least viscous?

■ How does the viscosity of foods influence how certain foods are used?

Figure 2–6 *A gas has no definite volume and will expand to fill its container. If allowed to, it will expand without limit. That is what happened to this balloon. A hole in Donald's arm allowed the helium gas within the balloon to escape into the atmosphere. Without gas, the arm hangs limply downward.*

Gases

Another phase of matter—the **gas** phase—does not have a definite shape or a definite volume. A gas fills all of the available space in a container, regardless of the size or shape of the container.

Although the particles of a gas tend to spread far out from one another, they can be pushed close together. When you pump air into a bicycle tire or blow up a party balloon, you squeeze a large amount of gas into a small volume. Fortunately, you can do this to the particles in a gas.

Just the opposite can also happen. The particles of a small amount of gas can spread out to fill a large volume. The smell of an apple pie baking in the oven in the kitchen will reach you in another room of your house because gases given off by the pie spread out to fill the whole house. In fact, if they are allowed to, gases will expand without limit. If not for the pull of gravity, the gases that make up the atmosphere of the Earth would expand into deep space. Can you explain then why a small planet like Mercury has little or no atmosphere?

Like liquids, gases have no definite shape. The particles that make up a gas are not arranged in any set pattern. So it is easy for gas particles to move around, either spreading apart or moving close together.

Figure 2–7 *Mercury, the closest planet to the sun, is a small planet. Because of its relatively small size, it does not have a great deal of gravity. What can you predict about Mercury's atmosphere?*

Figure 2–8 *A liquid has a definite volume but not a definite shape. It takes the shape of its container. An identical volume of liquid in three differently shaped glass vessels has three different shapes. A gas has neither a definite volume nor a definite shape. How would you describe the volume of a gas?*

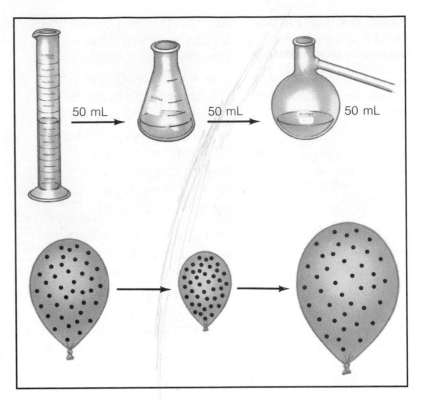

ACTIVITY

Determining Particle Space

1. Fill one 250-mL beaker with marbles, another with sand, and a third with water.

2. Describe the appearance of the beaker filled with marbles. Do the marbles occupy all the space in the beaker? Can you fit more marbles in the beaker?

3. Carefully pour some sand from its beaker into the beaker of marbles. How much sand are you able to pour? Is all the space in the beaker now occupied by marbles and sand?

4. Carefully add some water from its beaker to the beaker of marbles and sand. How much water can you add?

Is there space between the particles of a solid or a liquid?

■ How does what you observed partially explain the disappearance of rain on a lawn?

The behavior of gases can be explained in terms of the arrangement and movement of their particles. The world inside a container of gas particles is not as quiet as it may seem. Although you cannot see the particles of gas, they are in constant motion— moving about freely at speeds of nearly 500 meters per second. Whizzing around at such great speeds, the particles are constantly hitting one another. In fact, a single particle undergoes about 10 billion collisions per second! The particles are also colliding with the walls of the container. The effect of all these collisions is an outward pressure, or push, exerted by the gas. This pressure is what makes the gas expand to fill its container. What do you think would happen if the pressure in the container became too great?

BOYLE'S LAW Imagine you are holding an inflated balloon. If you press lightly on the outside of the balloon, you can feel the air inside pushing back. Now if you squeeze part of the balloon, what do you feel? You probably feel the air pressing against the wall of the balloon with even greater force.

This increase in pressure is due to a decrease in volume. By squeezing the balloon, you reduce the space the gas particles can occupy. As the particles are pushed a bit closer together, they collide with one another and with the walls of the balloon even more. So the pressure from the moving gas particles increases. The relationship between volume and pressure is explained by Boyle's law. According to Boyle's law, the volume of a fixed amount of gas varies inversely with the pressure of the gas. In other words, as one increases, the other decreases. If the volume of a gas decreases, its pressure increases. If the volume increases, its pressure decreases.

CHARLES'S LAW   Imagine that you still have that inflated balloon. This time you heat it very gently. What do you think happens to the volume of gas inside the balloon? As the temperature increases, the gas particles absorb more heat energy. They speed up and move farther away from one another. So the increase in temperature causes an increase in volume. If the temperature had decreased, the volume would have decreased. This relationship between temperature and volume is explained by Charles's law. According to Charles's law, the volume of a

Figure 2–9 In a solid, such as these crystals of iron pyrite, or fool's gold (top left), the particles are packed closely together and cannot move far out of place. In a liquid, such as molten iron (top center), the particles are close together but are free to move about or flow. In a gas, such as iodine, bromine, and chlorine (top right), the particles are free to spread out and fill the available volume.

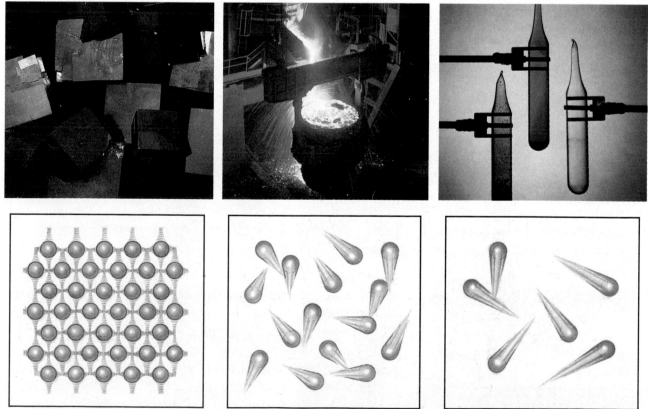

SOLID   LIQUID   GAS

ACTIVITY DOING

Demonstrating Charles's Law

1. Inflate a balloon. Make sure that it is not so large that it will break easily. Make a knot in the end of the balloon so that the air cannot escape.

2. Measure and record the circumference of the balloon. You can measure the circumference by placing a string around the fattest part of the balloon. Place your finger at the spot where one end of the string touches another part of the string. Now use a ruler to measure the distance between the two spots.

3. Place the balloon in an oven set at a low temperature—not more than 65°C (150°F). Leave the balloon in the oven for about 5 minutes.

4. Remove the balloon and quickly use the piece of string to measure its circumference. Record this measurement.

5. Now place the balloon in a refrigerator for 15 minutes.

6. Remove the balloon and immediately measure and record its circumference.

What happens to the size of the balloon at the higher temperature? At the lower temperature? Do the results of your investigation support Charles's law?

fixed amount of gas varies directly with the temperature of the gas. If the temperature of a gas increases, its volume increases. What do you think happens if the temperature decreases?

Boyle's law and Charles's law together are called the gas laws. The gas laws describe the behavior of gases with changes in pressure, temperature, and volume.

Plasma

The fourth phase of matter is called **plasma.** Plasma is quite rare on Earth. But the plasma phase is actually one of the most common phases in which matter is found in the universe. For example, stars such as the sun contain matter in the plasma phase.

Matter in the plasma phase is extremely high in energy and therefore dangerous to living things. Plasma can be made on the Earth only by using equipment that produces very high energy. But plasma cannot be contained by the walls of a container

Figure 2–10 *The illustration at top shows the effect of increasing the pressure on a fixed amount of gas. If the pressure is doubled, the volume is halved. The illustration at bottom shows the effect of increasing the temperature on the volume of a fixed amount of gas. If the temperature is doubled, the volume also doubles. What do you think happens to the volume of a fixed amount of gas when the pressure is halved? When the temperature decreases by half?*

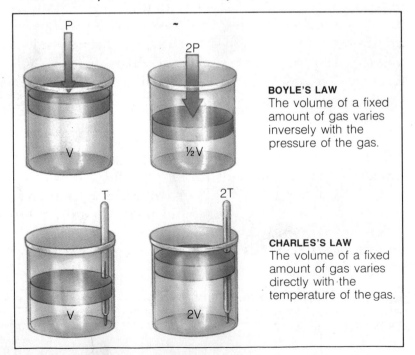

BOYLE'S LAW
The volume of a fixed amount of gas varies inversely with the pressure of the gas.

CHARLES'S LAW
The volume of a fixed amount of gas varies directly with the temperature of the gas.

made of ordinary matter, which it would immediately destroy. Instead, magnetic fields produced by powerful magnets are needed to keep the high-energy plasma from escaping. One day, producing plasmas on the Earth may meet most of our energy needs.

Figure 2–11 *Because the sun is a ball of matter with tremendous energy, matter exists there in the plasma phase. This photograph shows a huge solar flare erupting from the sun's surface.*

2-1 Section Review

1. What are the four phases of matter?
2. How is a crystalline solid different from an amorphous solid?

Critical Thinking—*Making Predictions*

3. Using the gas laws, predict what will happen to the volume of a gas if (a) the pressure triples, (b) the temperature is halved, (c) the pressure is decreased by a factor of five, (d) the pressure is halved and the temperature is doubled.

2-2 Phase Changes

Earth has been called the "water" planet. It is because of this abundant water that life can exist on Earth. But did you know that the water that makes up almost three fourths of the Earth's surface and about 1 percent of its atmosphere exists in three different phases? Ice, liquid water, and water vapor are all the same substance. What, then, causes the particles of a substance to be in one particular phase rather than another? The answer has to do with energy—energy that can cause the particles in a substance to move faster and farther apart.

A solid substance tends to have less energy than that same substance in the liquid phase. A gas usually has more energy than the liquid phase of the same substance. So ice has less energy than liquid

Guide for Reading

Focus on this question as you read.

▶ *What is the relationship between energy and phase changes?*

Figure 2–12 *Matter can change phase when energy is added or taken away. Is energy being added to or taken away from this ice cube?*

Melting Ice and Freezing Water

1. Place several ice cubes, a little water, and a thermometer in a glass. Wait several minutes.

2. Observe and record the temperature as the ice melts. This temperature is the melting point of ice.

3. Place a glass of water with a thermometer in it in a freezer. Observe and record the temperature every 10 minutes. Record the temperature of the water when it begins to freeze. This temperature is the freezing point of water. **Note:** *Do not let the glass of water freeze completely.*

How does the melting point of ice compare with the freezing point of water?

■ Plan an investigation to determine the effect of antifreeze on the freezing point of water.

water, and steam has more energy than ice or liquid water. The greater energy content of steam is what makes a burn caused by steam more serious than a burn caused by hot water!

Because energy content is responsible for the different phases of matter, substances can be made to change phase by adding or taking away energy. The easiest way to do this is to heat or cool the substance. This allows heat energy to flow into or out of the substance. This idea should sound familiar to you since you frequently increase or decrease heat energy to produce phase changes in water. For example, you put liquid water into the freezer to remove heat and make ice. And on a stove you add heat to make liquid water turn to steam.

The phase changes in matter are melting, freezing, vaporization, condensation, and sublimation. Changes of phase are examples of physical changes. In a physical change, a substance changes from one form to another, but it remains the same kind of substance. No new or different kinds of matter are formed, even though physical properties may change.

Solid-Liquid Phase Changes

What happens to ice cream on a hot day if you do not eat it quickly enough? It begins to melt. **Melting** is the change of a solid to a liquid. Melting occurs when a substance absorbs heat energy. The rigid crystal structure of the particles breaks down, and the particles are free to flow around one another.

The temperature at which a solid changes to a liquid is called the **melting point.** Most substances have a characteristic melting point. It is a physical

Figure 2–13 *If the children had eaten their ice cream cones quickly, the ice cream would have remained a solid until it was eaten. But because they took their time, heat energy caused the solid ice cream to become a liquid.*

LIFE SCIENCE LIBRARY/WATER. Photography by Ken Kay. Time-Life Books, Inc. Publisher ©1986 Time, Inc.

Figure 2–14 *Energy from within the Earth is great enough to melt rocks. Now in the liquid phase, the melted rocks flow from a volcano as a stream of lava.*

property that helps to identify the substance. For example, the melting point of ice is 0°C. The melting point of table salt is 801°C, whereas the melting point of a diamond is 3700°C.

The opposite phase change—that of a liquid changing to a solid—is called **freezing.** Freezing occurs when a substance loses heat energy. The temperature at which a substance changes from a liquid to a solid is called the **freezing point.** Strangely enough, the freezing point of a substance is equal to its melting point. So ice melts at 0°C and water freezes at 0°C.

Substances called alcohols have freezing points much lower than 0°C. Because of this property, these substances have an important use: They are used in automobile antifreeze. When alcohols are added to the water in an automobile's radiator, they lower the freezing point of the mixture. So even the coldest winter temperatures will not cause the water in the radiator to freeze. One such alcohol, ethylene glycol, when mixed with water can lower the freezing point of the mixture to −49°C.

The fact that freezing involves a loss of heat energy explains the "magic" worked by the orange growers you read about at the beginning of this chapter. The liquid water sprayed onto the trees released heat energy as it froze. Some of this heat energy was released into the oranges, keeping them from freezing.

Figure 2–15 *Would you believe that freezing water can produce a violent explosion? A cast-iron ball filled with water is placed in a beaker of dry ice and alcohol. As the water freezes and expands, a huge amount of force is exerted against the walls of the cast-iron ball, causing it to eventually explode.*

Liquid-Gas Phase Changes

Have you ever left a glass of water standing on the kitchen counter overnight? If so, did you notice that the water level was lower the next morning? Some of the liquid in the glass changed phase and became a gas. The gas then escaped into the air.

The change of a substance from a liquid to a gas is called **vaporization** (vay-puhr-ih-ZAY-shuhn). During this process, particles in a liquid absorb enough heat energy to escape from the liquid phase. If vaporization takes place at the surface of the liquid, the process is called **evaporation** (ee-vap-uh-RAY-shuhn). So some of the water you left in the glass overnight evaporated.

Evaporation is often thought of as a cooling process. Does this sound strange to you? Think for a moment about perspiration on the surface of your skin. As the water in perspiration evaporates, it absorbs and carries away heat energy from your body. In this way, your body is cooled. Can you explain why it is important for you to sweat on a hot day or after you perform strenuous exercise?

Vaporization does not occur only at the surface of a liquid. If enough heat energy is supplied, particles inside the liquid can change to gas. These particles travel to the surface of the liquid and then into the air. This process is called **boiling.** The temperature at which a liquid boils is called its **boiling point.** The boiling point of water under normal conditions at sea level is 100°C. The boiling point of table salt is 1413°C, and that of a diamond is 4827°C!

The boiling point of a liquid is related to the pressure of the air above it. Since the gas particles must escape from the surface of the liquid, they

Figure 2–16 *During both evaporation (left) and boiling (right), particles of a liquid absorb heat energy and change from the liquid phase to the gas phase. Based on this illustration, what is the difference between evaporation and boiling?*

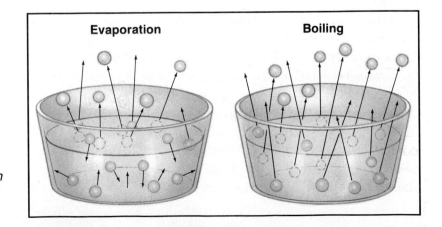

Figure 2–17 *The phase change from gas to liquid is called condensation. Water vapor in the air can condense and form rain. With the aid of cool night temperatures, water vapor in the air can also condense to form drops of dew.*

need to have enough "push" to equal the "push" of the air pressing down. So the lower the air pressure (the less the "push" of the air pressing down), the more easily the bubbles of gas can form within the liquid and then escape. Thus, lowering the air pressure lowers the boiling point.

At high altitudes, air pressure is much lower, and so the boiling point is reduced. If you could go many kilometers above the Earth's surface, the pressure of the air would be so low that you could boil water at ordinary room temperature! However, this boiling water would be cool. You would not be able to cook anything in this water. For it is the heat in boiling water that cooks food, not simply the boiling process.

Gases can change phase too. If a substance in the gas phase loses heat energy, it changes into a liquid. Scientists call this change in phase **condensation** (kahn-duhn-SAY-shuhn). You have probably noticed that cold objects, such as glasses of iced drinks, tend to become wet on the outside. Water vapor present in the surrounding air loses heat energy when it comes in contact with the cold glass. The water vapor condenses and becomes liquid drops on the glass. Can you think of another example of condensation?

ACTIVITY

WRITING

Some Fuelish Thoughts

Almost everyone in the United States depends upon the flammability of fuels to produce the energy needed to warm their homes and light their way. For the average user, these fuels are available in three phases: solids, liquids, and gases. Compile a list of several commonly used fuels and the phase in which they are used to produce energy. Then write a story that describes what would happen if all the fuels in one of the phases disappeared overnight.

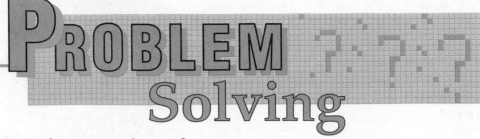

PROBLEM ??? Solving

It's Only a Passing Phase

Heat plays an important role in phase changes. Heat is energy that causes particles of matter to move faster and farther apart. As particles move faster, they leave one phase and enter another. Phase changes produce changes in only the physical properties of matter. They do not produce changes in the chemical properties. A substance is still the same kind of matter regardless of its phase.

The accompanying diagram is called a phase-change diagram. It shows the heat energy-temperature relationships as an ice cube becomes steam. Study the diagram and then answer the following questions.

Interpreting Diagrams

1. At which points does the addition of heat energy cause an increase in temperature?

2. At which points is there no temperature change despite the addition of heat energy?

3. What is happening at these points?

4. What is happening to the heat energy at the points where there is no temperature drop?

5. How can you apply this information to activities and/or occurrences in your daily life?

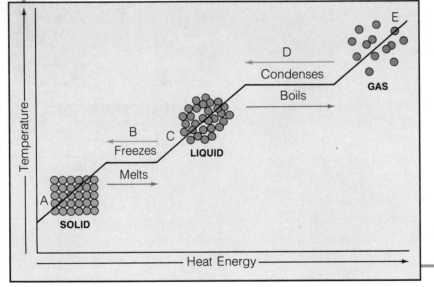

Solid-Gas Phase Changes

If you live in an area where winters are cold, you may have noticed something unusual about fallen snow. Even when the temperature stays below the melting point of the water that makes up the snow, the fallen snow slowly disappears. What happens to it? The snow undergoes **sublimation** (suhb-luh-MAY-shuhn). When a solid sublimes, its surface particles escape directly into the gas phase. They do not pass through the liquid phase.

A substance called dry ice is often used to keep other substances, such as ice cream, cold. Dry ice is solid carbon dioxide. At ordinary pressures, dry ice cannot exist in the liquid phase. So as it absorbs heat energy, it sublimes, or goes from the solid phase directly to the gas phase. By absorbing and carrying off heat energy as it sublimes, dry ice keeps materials that are near it cold and dry. Just think what would happen to an ice cream cake if it was packed with regular ice—ice that becomes liquid water before entering the gas phase—rather than with dry ice.

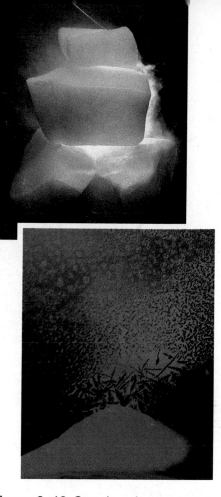

Figure 2–18 *Certain substances can go from the solid phase directly to the gas phase. Here you see dry ice becoming gaseous carbon dioxide (top) and iodine crystals becoming gaseous iodine (bottom). What is this process called?*

2–2 Section Review

1. How can substances be made to change phase?
2. What is a melting point? A freezing point?
3. What is the difference between evaporation and condensation?
4. Describe the changes in heat energy and particle arrangement as dry ice sublimes.

Critical Thinking—*Applying Concepts*

5. Suppose you place several ice cubes in a glass of water that is at room temperature. What happens to the ice cubes over time? What happens to the temperature of the water in the glass? What happens to the level of water in the glass? (Assume that you do not drink any of the liquid and that evaporation does not occur to any great extent.)

2–3 Chemical Properties and Changes

At the beginning of this chapter, you learned that you could identify different substances by comparing their physical properties. It was easy to see differences in color, shape, hardness, and volume in solid objects. But now suppose you have to distinguish between two gases: oxygen and hydrogen. Both are colorless, odorless, and tasteless. Since they are

Guide for Reading

Focus on this question as you read.

▶ *What is the difference between a chemical property and a chemical change?*

Figure 2–19 *Flammability is an important chemical property that may affect your life directly—as it does this firefighter, who watches helplessly as a home is totally destroyed by flames.*

gases, they have no definite shape or volume. And although each has a specific density, you cannot determine that density by dropping the gases into water to see what happens. In this particular case, physical properties are not very helpful in identifying the gases.

Fortunately, physical properties are not the only way to identify a substance. Both oxygen and hydrogen can turn into other substances and take on new identities. And the way in which they do this can be useful in identifying these two gases. The properties that describe how a substance changes into other new substances are called **chemical properties.**

In this case, if you collected some hydrogen in a test tube and put a glowing wooden stick in it, you would hear a loud pop. The pop occurs when hydrogen combines with oxygen in the air. What is actually happening is that the hydrogen is burning. The ability to burn is called **flammability** (flam-uh-BIHL-uh-tee). It is a chemical property. A new kind of matter forms as the hydrogen burns. Do you know what this new substance is? This substance—a combination of oxygen and hydrogen—is water.

Oxygen is not a flammable gas. It does not burn. But oxygen does support the burning of other substances. A glowing wooden splint placed in a test tube of oxygen will continue to burn until the oxygen is used up. This ability to support burning is another example of a chemical property. By using the chemical properties of flammability and supporting burning, you can distinguish between the two gases hydrogen and oxygen.

The changes that substances undergo when they turn into other substances are called **chemical changes.** Chemical changes are closely related to chemical properties, but they are not the same. **A chemical property describes a substance's ability to change into a different substance; a chemical change is the process by which the substance changes.** For example, the ability of a substance to burn is a chemical property. However, the process of burning is a chemical change. Figures 2–20 and 2–21 show several chemical changes.

Another name for a chemical change is a **chemical reaction.** Chemical reactions often involve chemically combining different substances. For example, during the burning of coal, oxygen combines chemically with carbon—the substance that makes up most of the coal. This combining reaction produces a new substance—carbon dioxide. The carbon and oxygen have changed chemically. They no longer exist in their original forms.

The ability to use and control chemical reactions is an important skill. For chemical reactions produce a range of products, from glass to pottery glazes to medicines. Your life is made easier and more enjoyable because of the products of chemical reactions.

Figure 2–20 *Nylon was one of the first synthetic fibers. Here you can see threads of nylon forming as chemicals squirted from barely visible holes undergo changes.*

Figure 2–21 *Many chemical changes occur in the world around you. Rust formed on the ship when iron combined with oxygen in the air. The copper in this statue reacted with sulfur in the air to form the soft green substance called verdigris. Fireworks produce beautiful colors and forms as a result of chemical changes. Chemical changes also occur in a leaf with the approach of cold winter weather.*

Synthetic fibers such as nylon and rayon, plastics, soaps, building materials, and even some of the foods you eat are the products of chemical reactions. The next time you eat a piece of cheese or a slice of bread, remember that you are eating the product of a chemical reaction.

2–3 Section Review

1. Give two examples of chemical properties.
2. What is chemical change? Give an example.

Connection—*Astronomy*

3. Suppose you visited another planet and wanted to test a sample of the planet's air. What kinds of tests would you perform to determine some physical properties of your sample? What kinds of tests would you perform to determine some chemical properties?

The Mess We Make

Throughout history, people have lived in groups: small family groups and larger groups such as those found in towns and cities. All people, in groups large and small, produce wastes. The amount of wastes each person produces is astounding—and, unfortunately, increasing! It has been estimated that in 1900 each person living in New York City produced 538 kilograms of waste. In 1989, the amount of waste produced by each person jumped to more than 825 kilograms. In a city of 8 million people, such amounts stagger the imagination and tax the ability of a city to deal with them.

Some of the waste materials—food scraps, paper, and other natural materials—are *biodegradable*. Biodegradable materials are capable of undergoing chemical changes that cause them to break down over time. Tiny animals, plants, and microscopic organisms such as bacteria that live in the soil are responsible for these changes. The result is that biodegradable materials are broken down into simpler chemical substances. Some of these chemical substances can be used by organisms for growth and repair. Others can be used as a source of energy. Biodegradable materials also undergo physical changes.

Certain waste materials, however, such as some plastics, are not biodegradable. These materials do not break down.

They remain intact in the environment for many hundreds of years. Scientists are now working to replace many of the non-biodegradable materials we use with those that are biodegradable. This would drastically reduce the amount of nonbiodegradable wastes we produce.

The problem of *waste disposal* is sure to loom ever greater in our future. The areas where we can safely dump wastes are rapidly filling up. Other methods to deal with waste materials will have to be developed soon in order to prevent the Earth from becoming a tremendous garbage dump tomorrow. You can help even now. You can use materials that are biodegradable. For example, you can use products wrapped in biodegradable materials. You can consume less. The less wastes you produce, the smaller the problem of waste disposal becomes. We must all assume responsibility for our actions—every little bit helps.

New York City, one of the world's largest cities, has the world's largest solid-waste dump site. In but a few more years, this site will be full—unable to accept another scrap of paper. It is imperative that we limit the amount of wastes we add to the environment.

Laboratory Investigation

Observing a Candle

Problem

How can physical and chemical properties be distinguished?

Materials (per group)

small candle
matches
metric ruler
candle holder or small empty food can
  with sand

Procedure

1. On a separate sheet of paper, prepare a data table similar to the one shown here.
2. Observe the unlighted candle. List as many physical and chemical properties as you can.
3. Place the candle in the candle holder. If you are not using a candle holder, fill the small food can with sand and place the candle in the center of the sand. Make sure that the candle is placed securely.
4. Under your teacher's supervision, carefully light the candle.
5. Observe the lighted candle. Continue to list as many physical and chemical properties as you can. Record your observations in the correct columns in your data table.

	Physical properties	Chemical properties
Unlighted candle		
Lighted candle		

Observations

1. What physical properties of the unlit candle did you observe?
2. What senses did you use when you made these observations?
3. What physical changes did you observe after you lit the candle?
4. What did you have to do to observe a chemical property of the candle?
5. What evidence of chemical change did you observe?

Analysis and Conclusions

1. What do you think is the basic difference between a physical property and a chemical property?
2. Can a physical property be observed without changing the substance?
3. What name is given to a process such as burning a candle? What is the result of such a process?
4. Which type of property—physical or chemical—is easier to determine? Why?
5. **On Your Own** Obtain a recipe for making bread. List the chemical and physical properties of the ingredients. How do the properties of the ingredients result in a loaf of bread?

Summarizing Key Concepts

2–1 Phases of Matter

▲ Physical properties of matter include color, shape, hardness, and density.

▲ A physical change occurs when the physical properties of a substance are altered. However, the substance remains the same kind of matter.

▲ Matter can exist in any of four phases: solid, liquid, gas, and plasma.

▲ A solid has a definite shape and volume.

▲ A crystal is the regular, repeating pattern in which the particles of some solids are arranged.

▲ Amorphous solids do not form crystals and thus do not keep a definite shape.

▲ A liquid has a definite volume but not a definite shape. A liquid takes the shape of its container.

▲ A gas has no definite shape or volume.

▲ Boyle's law states that the volume of a fixed amount of gas varies inversely with the pressure. Charles's law states that the volume of a fixed amount of gas varies directly with the temperature.

▲ Matter in the plasma state is very high in energy.

2–2 Phase Changes

▲ Phase changes are accompanied by either a loss or a gain of heat energy.

▲ Melting is the change of a solid to a liquid at a temperature called the melting point. Freezing is the change of a liquid to a solid at the freezing point.

▲ Vaporization is the change of a liquid to a gas. Vaporization at the surface of a liquid is called evaporation. Vaporization throughout a liquid is called boiling.

▲ The boiling point of a liquid is related to the air pressure above the liquid.

▲ The change of a gas to a liquid is called condensation.

▲ The change of a solid directly to a gas without going through the liquid phase is called sublimation.

2–3 Chemical Properties and Changes

▲ Chemical properties describe how a substance changes into a new substance.

▲ Flammability, the ability to burn, is a chemical property.

▲ When a substance undergoes a chemical change, or a chemical reaction, it turns into a new and different substance.

Reviewing Key Terms

Define each term in a complete sentence.

2–1 Phases of Matter
physical property
phase
solid
crystal
liquid
gas
plasma

2–2 Phase Changes
melting
melting point
freezing
freezing point
vaporization
evaporation
boiling
boiling point
condensation
sublimation

2–3 Chemical Properties and Changes
chemical property
flammability
chemical change
chemical reaction

Chapter Review

Content Review

Multiple Choice

Choose the letter of the answer that best completes each statement.

1. Color, odor, and density are
   a. chemical properties.
   b. chemical changes.
   c. physical properties.
   d. solid properties.
2. A regular pattern of particles is found in
   a. molecules.   c. compressions.
   b. crystals.   d. plasmas.
3. The phase of matter that is made up of very high-energy particles is
   a. liquid.   c. gas.
   b. plasma.   d. solid.
4. As the volume of a fixed amount of gas decreases, the pressure of the gas
   a. decreases.
   b. remains the same.
   c. first increases then decreases.
   d. increases.

5. As the temperature of a fixed amount of gas increases, the volume
   a. decreases.
   b. remains the same.
   c. increases then decreases.
   d. increases.
6. All liquids have
   a. definite shape and definite volume.
   b. no definite shape but definite volume.
   c. no definite shape and no definite volume.
   d. definite shape but no definite volume.
7. A solid changes to a liquid by
   a. evaporation.   c. melting.
   b. freezing.   d. sublimation.
8. Vaporization that takes place at the surface of a liquid is called
   a. boiling.   c. sublimation.
   b. evaporation.   d. condensation.

True or False

If the statement is true, write "true." If it is false, change the underlined word or words to make the statement true.

1. Particles that make up a solid are packed very close together.
2. The particles of matter are spread farthest apart in a liquid.
3. The relationship between the temperature of a gas and the volume it occupies is described by Boyle's law.
4. A liquid will freeze when it absorbs heat energy.
5. The process by which a liquid changes to a gas is called vaporization.
6. Drops of water on the outside of a cold glass are water vapor that has sublimed into a liquid.
7. New substances that have different properties are formed as a result of physical changes.

Concept Mapping

Complete the following concept map for Section 2–1. Refer to pages N6–N7 to construct a concept map for the entire chapter.

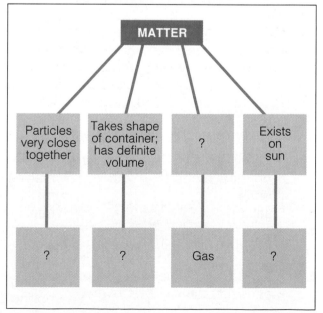

Concept Mastery

Discuss each of the following in a brief paragraph.

1. Identify the following properties as either physical or chemical. Explain your answers. (a) taste, (b) flammability, (c) color, (d) odor, (e) ability to dissolve, (f) tendency to rust

2. Identify the following changes as either physical or chemical. Explain your answers. (a) burning coal, (b) baking brownies, (c) digesting food, (d) dissolving sugar in hot water, (e) melting butter, (f) exploding fireworks, (g) rusting iron

3. Explain why both the melting point and the freezing point of water are 0°C.

4. Compare the solid, liquid, and gas phases of matter in terms of shape, volume, and arrangement and movement of molecules.

5. Explain how evaporation and boiling are similar. How are they different?

Critical Thinking and Problem Solving

Use the skills you have developed in this chapter to answer each of the following.

1. **Applying concepts** I am one of the most common substances on Earth. I am easily seen in the liquid phase and the solid phase. I am hard to observe as a gas. Identify me and explain the clues.

2. **Applying concepts** You are coming home from school one cold winter's day when you observe a neighbor filling the radiator of his car with plain tap water. What advice would you give this person? Why?

3. **Making inferences** Several campers who have set up camp on a high mountain peak decide they would like to enjoy a cup of coffee. They notice that although the water they use to make the coffee has boiled, the coffee does not seem as hot as the coffee they drink while camping out at sea level. Use your knowledge of the effects of altitude on boiling to explain their observation.

4. **Relating cause and effect** Rubbing alcohol, or isopropyl alcohol, evaporates quickly at room temperature. Explain why people with a high fever are often given rubdowns with isopropyl alcohol as a means of reducing their fevers.

5. **Applying concepts** An automobile mechanic may suggest that you test the pressure of the air in a car's tires after the car has been moving for a while and the tires have heated up. Why do you think this is good advice?

6. **Developing a hypothesis** Solid room air fresheners "disappear" over a period of time and must be replaced. What do you think happens to the solid? How is this related to the way in which the freshener releases its pleasant odor?

7. **Using the writing process** Write a 250-word story to describe the day on which your birthday party was almost a disaster. Use the following words in your story: boiling, freezing, crystal, evaporation, liquid, melting, phase, physical change, solid, sublimation.

Mixtures, Elements, and Compounds

3

Guide for Reading

After you read the following sections, you will be able to

3–1 Classes of Matter

■ Describe how matter is classified according to its makeup.

3–2 Mixtures

■ List the different kinds of mixtures.

■ Compare the properties of solutions with the properties of other mixtures.

3–3 Elements

■ Explain why elements are considered pure substances.

3–4 Compounds

■ Explain why compounds are considered pure substances.

■ Describe how chemical symbols, formulas, and balanced equations are used to describe a chemical reaction.

A reddish stain on a scrap of fabric . . . some bits of dust gathered in the creases of a man's clothing . . . a seemingly unimportant clump of mud in the corner of a room . . . a few pieces of pipe tobacco found at the murder scene . . . What could all these details mean? To detective Sherlock Holmes, the creation of British author Arthur Conan Doyle, they were clues to the most mysterious crimes imaginable. By paying close attention to the evidence, Holmes was able to solve many perplexing crimes. And in so doing, he amazed not only the London police but also his own assistant, Dr. Watson.

Holmes's success had a simple, solid basis: logical thinking combined with a knowledge of chemistry. Using this knowledge, he was able to classify and analyze various substances that were clues to the mysteries. Holmes was a master at using scientific principles to solve crimes.

The whole world is a place of mystery, filled with puzzles and wonders that await the investigation of detectives like you. But before you set out on your adventure, you will need to know how chemical substances are classified. And soon you will share the delight of Holmes, who exclaimed at moments of discovery, "By Jove, Watson, I've got it!"

Journal *Activity*

You and Your World Detectives often use the scientific method to solve crimes. Pretend that you are a detective who has been asked to solve the theft of cookies from a jar in the student cafeteria. There are many suspects. Write down the method you would use to solve this "terrible" crime.

◄ *Basil Rathbone played Sherlock Holmes in a series of movies that detailed the exploits of the great fictional detective.*

How to Watch the Foods You Eat, p.154

Figure 3–1 *It is much easier to select exactly what you want to purchase if articles are grouped together. The vegetables are classified by type; the wool, by color.*

3–1 Classes of Matter

Have you ever collected rocks, stamps, or marbles? If so, you probably know how important it is to classify, or group, the objects in a collection. To do this, you might use characteristics such as color, shape, or texture. Or maybe you would classify the objects in a collection according to their uses. In any case, you would be using a classification system based on a particular property to group the objects.

Classification systems are used all the time to organize objects. Books in a library are arranged in an organized manner. So too are the clothes in a department store and the food in a supermarket. Next time you are in a record store, notice how the records and tapes have been organized.

You can see how classifying objects—whether they be collections, books, foods, or tapes—makes it easier for you to organize them (and to locate a particular item). In order to make the study of matter easier to understand, scientists have developed different ways to classify matter. In Chapter 2, you learned that matter exists in four phases: solid, liquid, gas, and plasma. Phases are one way to classify matter.

But classifying matter by phase is not specific enough and can lead to confusion. One kind of substance can exist in more than one phase. Water is a good example. Water can be a solid in the form of ice, a liquid, or a gas in the form of water vapor. How, then, would you classify water?

Classifying matter according to phase often groups very different substances together. Table salt, gold, steel, and sand are all solids. Although they are all solids, they differ from each other in many important ways. Should they be grouped together? What about water and gasoline, which are both clear liquids? In what ways do these two liquids differ?

In order to make the study of matter easier, scientists have used a classification system based on the makeup of matter. **According to makeup, matter exists as mixtures, solutions, elements, or compounds.**

Figure 3–2 *Although they are both clear liquids, gasoline and water differ from each other in important ways. Would this truck run well if water was put into the tank instead of gasoline or diesel fuel?*

3–1 Section Review

1. According to makeup, what are the four classes of matter?
2. Why is it more useful to classify matter according to makeup rather than according to phase?

Critical Thinking—*Applying Concepts*

3. A librarian wanted to save space, so he decided to classify books according to size. By putting all the small books together and all the large books together, he was able to fit more shelves in a bookcase. Soon, the number of readers using the library decreased dramatically. Why was this method of classification not appreciated by the readers?

ACTIVITY

Classifying Common Objects

1. Obtain samples of the following materials: sugar, salt water, copper wire, taco shell, pencil eraser.

2. Observe each material. Describe its appearance in a few sentences.

3. Use simple physical tests to determine which of your samples are mixtures, solutions, elements, or compounds.

4. Present your observations in a chart.

3-2 Mixtures

Look at the photograph of a piece of granite in Figure 3–3. Granite, which is a type of rock, is made of different minerals mixed together. You can see some of these minerals—quartz, mica, and feldspar—when you look at the granite. Sand is also made of different minerals mixed together. When you pick up a handful of sand, you see dark and light grains mixed together. Granite, sand, soil, concrete, and salad dressing are examples of matter that consists of several substances mixed together.

Matter that consists of two or more substances mixed together but not chemically combined is called a mixture. A **mixture** is a combination of substances. Each substance that makes up a mixture has its own specific properties and is the same throughout. But the mixture as a whole is not the same throughout. Let's go back to our example of granite. As you just read, granite is a mixture of minerals. The individual minerals in granite share the same properties. Every piece of quartz has the same properties as every other piece of quartz. This is true of mica and feldspar also.

Properties of Mixtures

The substances in a mixture are not chemically combined. The substances keep their separate identities and most of their own properties. This is an important property of mixtures. Think for a moment of a mixture of sugar and water. When sugar and water are mixed, the water is still a colorless liquid. The sugar still keeps its property of sweetness even though it is dissolved in the water. Although they may look identical, you can easily taste the difference between plain water and a sugar-water mixture.

Substances in a mixture may change in physical appearance, as when they dissolve. Some physical properties of the mixture, such as its melting point and boiling point, may also change. But the substances do not change in chemical composition. In the sugar-water mixture, the same particles of water and sugar are present after the mixing as before it. No new chemical substances have been formed.

Figure 3–3 *Granite rock is made of the minerals quartz, feldspar, and mica. It is not a pure substance. Is granite a mixture? Why?*

If you eat cereal for breakfast, you are probably making a mixture. That is what you produce when you pour milk over the cereal. And if you put berries, banana slices, or raisins into your cereal, you make an even more complex mixture. But you do not use exactly the same amounts of cereal, milk, and fruit each time. This illustrates another property of mixtures.

The substances that make up a mixture can be present in any amount. The amounts are not fixed. A salt-and-pepper mixture, for example, may be one-third salt and two-thirds pepper, or one-half salt and one-half pepper. You can mix lots of sugar or only a little in a glass of iced tea. But in both cases, the mixture is still iced tea.

Because the substances in a mixture retain their original properties, they can be separated out by simple physical means. Look at Figure 3–4. A mixture of powdered iron and powdered sulfur has been made. The particles of iron are black, and the particles of sulfur are yellow. The mixture of the two has a grayish color. If you look closely at the mixture, however, you will see that particles of iron and particles of sulfur are clearly visible. You can separate the particles that make up this mixture in a rather simple way. Because iron is attracted to a magnet and sulfur is not, iron can be separated from the sulfur by holding a strong magnet near the mixture. The particles of iron can be removed from the mixture by simple

Figure 3–4 *By combining powdered iron (top) with powdered sulfur (center), an iron-sulfur mixture is formed. What physical property of iron is being used to separate the mixture (bottom)?*

Figure 3–5 *Black sand on a Hawaiian beach, a superburger, and a nebula far from Earth are all mixtures. What are some properties of mixtures?*

What's This in My Food?

1. In a plastic sandwich bag, place half a cup of an iron-fortified breakfast cereal.

2. Squeeze the air from the bag. Seal the bag. Use your hands to crush the cereal into a fine powder.

3. Now open the bag and pour the crushed cereal into a bowl. Add just enough water to completely cover the cereal.

4. Stir the water and cereal mixture with a bar magnet for at least ten minutes.

5. Remove the magnet. Let the liquid on the magnet drain back into the bowl.

6. Use a piece of white tissue paper to remove the particles attached to the magnet. Use a hand lens to observe the particles.

What did you remove from the mixture? Is the cereal a heterogeneous or homogeneous mixture? Why?

physical means—in this case, a magnet can be used to separate the iron from the sulfur.

The methods used to separate substances in a mixture are based on the physical properties of the substances that make up the mixture. No chemical reactions are involved. What physical property of iron made it possible to separate it from sulfur in the sulfur-iron mixture? What are some other physical properties that can be used to separate the substances that make up a mixture? For example, how could you separate a mixture of sugar and water?

Types of Mixtures

You now know that granite is a mixture. Other mixtures include concrete and stainless steel. Concrete is a mixture of pieces of rock, sand, and cement. Stainless steel is a mixture of chromium and iron. (You might be interested to learn that stainless steel does "stain," or rust. However, it "stains less" than regular steel—hence, its name.) From your experience, you would probably say that stainless steel seems "better mixed" than concrete. You cannot see individual particles of chromium and iron in stainless steel, but particles of rock, sand, and cement are visible in concrete. Mixtures are classified according to how "well mixed" they are.

Figure 3–6 *This gold miner in Finland is separating heavy pieces of gold from lighter pieces of rock, sand, and soil by swirling the mixture in a shallow pan of water. The gold will settle to the bottom of the pan. Salt water is a mixture of salts and water. When the water evaporates, deposits of salt, such as these in Mono Lake, California, are left behind.*

HETEROGENEOUS MIXTURE A mixture that does not appear to be the same throughout is said to be heterogeneous. A **heterogeneous** (heht-er-oh-JEE-nee-uhs) **mixture** is the "least mixed" of mixtures. The particles in a heterogeneous mixture are large enough to be seen and to separate from the mixture. Concrete is an example of a heterogeneous mixture. Can you think of some other examples?

Not all heterogeneous mixtures contain only solid particles (as does concrete). Shake up some pebbles or sand in water to make a liquid-solid mixture. This mixture is easily separated just by letting it stand. The pebbles and sand will settle to the bottom of the jar. Oil and vinegar, often used as a salad dressing, make up a liquid-liquid heterogeneous mixture. When the mixture has been shaken well, drops of oil are spread throughout the vinegar. This mixture, too, will separate when allowed to stand. Now you know why you have to shake a bottle of salad dressing that has not been used for a while.

HOMOGENEOUS MIXTURE A mixture that appears to be the same throughout is said to be homogeneous (hoh-moh-JEE-nee-uhs). A **homogeneous mixture** is "well mixed." The particles that make up the mixture are very small and not easily recognizable. These particles do not settle when the mixture is allowed to stand. Stainless steel is a homogeneous mixture.

Although you may not be aware of it, many of the materials you use and eat every day are homogeneous mixtures. Milk, whipped cream, toothpaste, and suntan lotion are just a few examples. In these homogeneous mixtures, the particles are mixed together but not dissolved. As a group, these mixtures are called **colloids** (KAHL-oidz).

The particles in a colloid are relatively large in size and are kept permanently suspended. They are also continuously bombarded by other particles. This bombardment accounts for two important properties of a colloid. One property is that a colloid will not separate upon standing (as do many heterogeneous mixtures). Because the particles are constantly bombarded, they do not have a chance to settle out. Another property is that a colloid often appears cloudy. This is because the constant bombardment of particles

Figure 3–7 *Tacos are heterogeneous mixtures in which the parts are easy to recognize and to separate from the mixture.*

ACTIVITY

Danger in the Air

Fog, a colloid you are probably familiar with, forms when tiny droplets of water become suspended in air. Drive along a fog-shrouded road and you will soon realize how dangerous fog can be. On a clear day, you can see forever; on a foggy day, however, your car's headlights offer scant help in piercing the gloom.

You might like to find out about the poet Carl Sandburg's impressions of this common colloid by reading his poem "Fog."

Figure 3–8 *A gelatin desert is a colloid that contains liquid particles mixed with a solid. Whipped cream is a colloid that contains gas particles in a liquid. The smoke from a campfire is a colloid that contains solid particles mixed in a gas.*

enables a colloid to scatter light. So if a beam of light is passed through a colloid, the beam becomes visible. The white cloudy appearance of milk is due to the scattering of light by the bombarding particles in this familiar colloid. If you have ever seen a searchlight sweep through the air at night, you have observed another example of this property of colloids. Figure 3–9 is a table of several different types of colloids. Which ones are you familiar with?

Solutions

A solution (suh-LOO-shuhn) is a type of homogeneous mixture formed when one substance dissolves in another. You might say that a **solution** is the "best mixed" of all mixtures. You are probably familiar with many different solutions. Ocean water is one example. In this solution, different salts are dissolved in water. Another example of a solution is antifreeze. Lemonade and tea are also solutions. One important solution helps keep you alive. Do you know what it is? Air is a solution of oxygen and other gases dissolved in nitrogen.

All solutions have several important properties. Picture a glass of lemonade as you read about them. How do you think the lemonade was made? Lemon juice and sugar were probably added to water. They dissolved in the water. A solution always has a substance that is dissolved and a substance that does the dissolving. The substance that is dissolved is called the **solute** (SAHL-yoot). The substance that does the dissolving is called the **solvent** (SAHL-vuhnt). In the case of the lemonade, there are two solutes: lemon juice and sugar. The solvent is water.

Figure 3–9 *You might be surprised to learn how many commonly used materials are colloids. What type of colloid is mayonnaise? Butter?*

TYPES OF COLLOIDS	
Name	**Example**
Fog (liquid in gas)	Clouds
Smoke (solid in gas)	Smoke
Foam (gas in liquid)	Whipped cream
Emulsion (liquid in liquid)	Mayonnaise
Sol (solid in liquid)	Paint
Gel (liquid in solid)	Butter

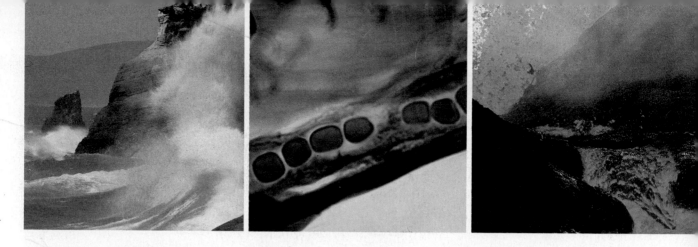

Looking at the glass of lemonade, you will notice that the particles are not large enough to be seen. Because the particles in a solution are so small, most solutions cannot easily be separated by simple physical means. Unlike many colloids, liquid solutions appear clear and transparent. The particles in a liquid solution are too small to scatter light.

Tasting the lemonade illustrates another property of a solution. Every part of the solution tastes the same. This might lead you to believe that one property of a solution is that its particles are evenly spread out. And you would be right!

There are nine possible types of solutions, as you can see from Figure 3–12 on page 66. Many liquid solutions contain water as the solvent. Ocean water is basically a water solution that contains many salts. Body fluids are also water solutions. Because water can dissolve many substances, it is called the "universal solvent."

Figure 3–10 *Solutions are the "best mixed" of all mixtures. Seawater, blood, and lava from an erupting volcano are all solutions. What other common solutions can you name?*

Activity Bank

Acid Rain Takes Toll on Art, p.156

Figure 3–11 *You can form three kinds of solutions using a liquid solvent. Name a common solution that is a solid dissolved in a liquid. A liquid dissolved in a liquid. A gas dissolved in a liquid.*

Solid solute

Liquid solute

Gas solute

Liquid solvent (water)

Liquid solvent (water)

Liquid solvent (water)

Figure 3–12 *Nine different types of solutions can be made from three phases of matter. Can most solutions be separated by simple physical means? Explain.*

TYPES OF SOLUTIONS		
Solute	**Solvent**	**Example**
Gas Gas Gas	Gas Liquid Solid	Air (oxygen in nitrogen) Soda water (carbon dioxide in water) Charcoal gas mask (poisonous gases on carbon)
Liquid Liquid Liquid	Gas Liquid Solid	Humid air (water in air) Antifreeze (ethylene glycol in water) Dental filling (mercury in silver)
Solid Solid Solid	Gas Liquid Solid	Soot in air (carbon in air) Ocean water (salt in water) Gold jewelry (copper in gold)

ACTIVITY
DISCOVERING

Is It a Solution?

1. Obtain samples of the following substances: sugar, flour, powdered drink, cornstarch, instant coffee, talcum powder, soap powder, gelatin.

2. Crush each substance into pieces of equal size. Make sure you keep the substances separate.

3. Determine how much of each substance you can dissolve in samples of the same amount of water at the same temperature. Determine how quickly each substance dissolves.

4. Determine which substances dissolved fastest and to the greatest extent. Record your findings in a data table.

■ Use your knowledge of the properties of solutions to determine which substances formed true solutions.

SOLUBILITY A substance that dissolves in water is said to be **soluble** (SAHL-yoo-buhl). Salt and sugar are soluble substances. Mercury and oil do not dissolve in water. They are **insoluble.**

The amount of a solute that can be completely dissolved in a given amount of solvent at a specific temperature is called its **solubility.** What is the relationship between temperature and the solubility of solid solutes? In general, as the temperature of a solvent increases, the solubility of the solute increases. What about gaseous solutes? An increase in the temperature of the solvent usually decreases the solubility of a gaseous solute. This explains why soda that warms up goes flat. The "fizz" of soda is due to bubbles of carbon dioxide dissolved in the solution. As soda warms, the dissolved CO_2 comes out of solution. It is less soluble. Without the CO_2, the soda tastes flat.

Some substances are not very soluble in water. But they dissolve easily in other solvents. For example, one of the reasons you use soap to wash dirt and grease from your skin or clothing is that soap dissolves these substances, whereas water alone does not. The soap dissolves the grease and then, along with the grease, it is washed away by the water.

ALLOYS Not all solutions are liquids. Solutions can exist in any of the three phases—solid, liquid, or gas—as the table in Figure 3–12 indicates. Metal solutions called **alloys** are examples of solids dissolved in solids. Gold jewelry is actually a solid solution of gold and copper. Brass is an alloy of copper and zinc.

Figure 3–13 *The important alloy stainless steel is a mixture of iron and chromium. Here it is being poured as a white-hot liquid from a vat. What type of mixture is stainless steel?*

Sterling silver contains small amounts of copper in solution with silver. And stainless steel, which you read about before, is an alloy of chromium and iron. You may find it interesting to learn about the make-up of other alloys, such as pewter, bronze, and solder. How do you think alloys are made?

3–2 Section Review

1. What is a mixture? What are three properties of a mixture?
2. How does a heterogeneous mixture differ from a homogeneous mixture?
3. What is a colloid?
4. What is a solution? What are the two parts of a solution? What are two properties of a solution?

Connection—*You and Your World*

5. Trout are fish that need to live in water that has a great deal of oxygen dissolved in it. What can you predict about the temperature of the water in a trout stream? Explain your answer. The correct answer may improve your luck the next time you go fishing!

ACTIVITY

DISCOVERING

Where's the Fizz?

You can determine what conditions affect the solubility of a gas in a liquid.

1. Remove the cap from a bottle of soda.

2. Immediately fit the opening of a balloon over the top of the bottle. Shake the bottle several times. Note any changes in the balloon.

3. Heat the bottle of soda very gently by placing it in a pan of hot water. Note any further changes in the balloon.

What two conditions of solubility are being tested here?

■ What general statement about the solubility of a gas in a liquid can you now make?

PROBLEM Solving

What's the Solution?

Look closely at the four photographs below. Each shows a familiar solution. For each photograph, tell what solute is dissolved in what solvent. List other solutions that are similar to the solutions shown. Use Figure 3–12 to help you. Solve this problem and you are part of the solution!

Guide for Reading

Focus on this question as you read.

▶ *What are the simplest pure substances?*

3–3 Elements

Scientists often examine, in great detail, the particles that make up a substance. Close observation shows that in some cases all the particles that make up a substance are alike; in other cases, all the particles are not alike. When all the particles are alike, the substance is called a **pure substance.** A pure substance is made of only one kind of material and has definite properties. A pure substance is the same throughout. All the particles in a pure substance are exactly the same. Iron, aluminum, water, sugar, and table salt are examples of pure substances. So is the oxygen your body uses from the air you breathe. A sample taken from any of these substances is identical to any other sample taken from that substance. For instance, a drop of pure water taken from a well in Arizona, a river in Australia, or the hard-packed snow and ice of Antarctica is the same.

Elements are the simplest pure substance. An **element** cannot be changed into a simpler substance by heating or by any chemical process. The particles that make up an element are in their simplest form. (You will learn just what these particles are in the next paragraph.) Suppose you melt a piece of iron by adding heat energy to it. You may think that you have changed the iron into a simpler substance. But the liquid iron you now have still contains only iron particles. True, the heat has changed the iron's phase—from a solid to a liquid. But it is still iron. No new or simpler substance has been formed.

Elements and Atoms

The smallest particle of an element that has the properties of that element is called an **atom.** An atom is the basic building block of matter. All elements are made of atoms. Atoms of the same element are alike. Atoms of different elements are different.

Scientists now know that an atom is made of even smaller particles. These particles, however, do not have the properties of the elements they make up. You will learn more about the particles that make up atoms in Chapter 4.

Chemical Symbols

For many years, scientists had to spell out the full names of elements when writing about them. As you can imagine, this practice was time consuming. Then in 1813, a system of representing the elements with symbols was introduced. After all, why couldn't chemists do what mathematicians and musicians had been doing for years?

ACTIVITY
DOING

Name That Element

From around your home, collect several items that are made of common elements. For example, items made of iron, copper, aluminum, or carbon should be fairly easy to find. See if you can find items that are made of less common elements.

Make a display of the elements you collect. Label each element with its name and chemical symbol.

Figure 3–14 *This miner examines a rock specimen closely because if he is lucky, the rocks he mines will contain the element gold.*

Figure 3–15 *Elements are the simplest type of pure substance. These yellow crystals are made of the element sulfur. Computer chips contain silicon. In what ways would our lives be different if there was no silicon on Earth?*

Chemical symbols are a shorthand way of representing the elements. Each symbol consists of one or two letters, usually taken from the element's name. The symbol for the element oxygen is O. The symbol for hydrogen is H; for carbon, C. The symbol for aluminum is Al; and for chlorine, Cl. Two letters are needed for a chemical symbol when the first letter of that element's name has already been used as a symbol for another element. For example, the symbol for carbon is C, for calcium it is Ca, and for copper it is Cu. You should note that when two letters are used in a symbol, the first letter is always capitalized but the second letter is never capitalized.

What do you think the symbol for gold is? The symbol for gold is Au. Does that surprise you? Gold is not spelled with an "a" or a "u." But the reason for the symbol is really not so strange. The Latin

Figure 3–16 *This table shows the chemical symbols for some of the most common elements. What is the symbol for tin? Why is Fe the symbol for iron?*

COMMON ELEMENTS

Name	Symbol	Name	Symbol	Name	Symbol
Aluminum	Al	Hydrogen	H	Oxygen	O
Bromine	Br	Iodine	I	Phosphorus	P
Calcium	Ca	Iron	Fe	Potassium	K
Carbon	C	Lead	Pb	Silicon	Si
Chlorine	Cl	Lithium	Li	Silver	Ag
Chromium	Cr	Magnesium	Mg	Sodium	Na
Copper	Cu	Mercury	Hg	Sulfur	S
Fluorine	F	Neon	Ne	Tin	Sn
Gold	Au	Nickel	Ni	Uranium	U
Helium	He	Nitrogen	N	Zinc	Zn

Figure 3–17 *The top four rows are symbols that were used by ancient alchemists to represent elements. The bottom three rows are part of the system developed by John Dalton. Do you think it would be easier to remember these symbols for the elements or the symbols shown in Figure 3–16?*

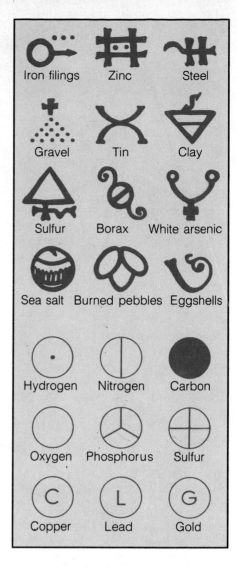

name for gold is *aurum.* Scientists often use the Latin name of an element as its symbol. Here are some other examples. The symbol used for silver is Ag, from the Latin word for silver, *argentum.* The Latin word for iron is *ferrum,* and the symbol for this element is Fe. The symbol for mercury is Hg, from the Latin name *hydrargyrum.* The table in Figure 3–16 lists some common elements and their symbols.

3–3 Section Review

1. What is a pure substance? Why are elements pure substances?
2. What is an atom? How do atoms of the same element compare? Of different elements?
3. Write the chemical symbols for oxygen, nitrogen, lead, sulfur, sodium, and helium.

Critical Thinking—*Relating Facts*
4. Why can elements be thought of as homogeneous matter?

3–4 Compounds

As you just learned, the simplest type of pure substance is an element. But not all pure substances are elements. Water and table salt, for example, are pure substances. Each is made of only one kind of material with definite properties. Yet water and table salt are not elements. Why? They can be broken down into simpler substances. Water can be broken down into the elements hydrogen and oxygen. Table salt can be broken down into the elements sodium and chlorine. Thus water and table salt, like many other pure substances, are made of more than one element.

Guide for Reading

Focus on this question as you read.

▶ *What is a compound?*

Pure substances that are made of more than one element are called compounds. A **compound** is two or more elements chemically combined. Sugar is a compound that is made of the elements carbon, hydrogen, and oxygen. Carbon dioxide, ammonia, baking soda, and TNT are compounds. Can you name some other common compounds?

Unlike elements, compounds can be broken down into simpler substances. Heating is one way to separate some compounds into their elements. The compound copper sulfide, which is also known as the ore chalcocite, can be separated into the elements copper and sulfur by heating it to a high temperature.

Electric energy is often used to break down compounds that do not separate upon heating. By passing an electric current through water, the elements hydrogen and oxygen can be obtained. What elements would you obtain if you passed an electric current through melted table salt?

The properties of the elements that make up a compound are often quite different from the properties of the compound itself. Would you want to flavor your French-fried potatoes with a poisonous gas and a highly active metal? Yet, in a way, this is exactly what you are doing when you sprinkle salt on your potatoes. Chlorine is a yellow-green gas that is poisonous. Sodium is a silvery metal that explodes if placed in water. But when chemically combined, these elements produce a white compound—sodium chloride—that you cannot and probably would not want to live without. And it adds a tasty flavor to foods as well!

Figure 3–18 *The element sodium is often stored under kerosene because it reacts explosively when it comes into contact with water (left). The element chlorine is a poisonous gas (center). When sodium and chlorine combine, sodium chloride forms. What is the common name for sodium chloride?*

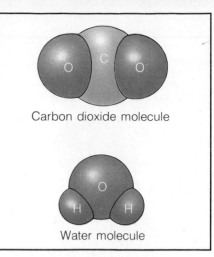

Carbon dioxide molecule

Water molecule

Figure 3–19 *As you can see from this diagram, molecules are made of two or more elements chemically bonded together. What are the chemical formulas for these two compounds?*

Compounds and Molecules

Most compounds are made of **molecules** (MAHL-ih-kyoolz). A molecule is made of two or more atoms chemically bonded together. A molecule is the smallest particle of a compound that has all the properties of that compound.

Water is a compound. A molecule of water is made up of 2 atoms of hydrogen chemically bonded to 1 atom of oxygen. One molecule of water has all of the properties of a glass of water, a bucket of water, or a pool of water. If a molecule of water were broken down into atoms of its elements, would the atoms have the same properties as the molecule?

Just as all atoms of a certain element are alike, all molecules of a compound are alike. Each molecule of ammonia, for example, is like every other. Because it is made of only one kind of molecule, a compound is the same throughout. So compounds, like elements, are pure substances.

Chemical Formulas

You probably learned the alphabet before you learned to read. Well, you can think of chemical symbols as the letters of a chemical alphabet. Just as you learned to put letters together to make words, chemical symbols can be put together to make chemical "words." Combinations of chemical symbols are called **chemical formulas.** Chemical formulas are a shorthand way of representing chemical substances.

Most chemical formulas represent compounds. For example, ammonia is a compound made of the elements nitrogen (N) and hydrogen (H). The chemical formula for ammonia is NH_3. A molecule of ammonia contains 1 atom of nitrogen and 3 atoms of hydrogen. The formula for rubbing alcohol is C_3H_7OH. What elements make up this compound? How about the compound silver nitrate, $AgNO_3$?

Sometimes a formula represents a molecule of an element, not a compound. For example, the symbol

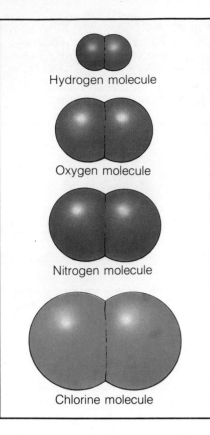

Hydrogen molecule

Oxygen molecule

Nitrogen molecule

Chlorine molecule

Figure 3-20 *Some elements are found in nature as molecules that are made of 2 atoms of that element. Hydrogen, oxygen, nitrogen, and chlorine are examples of such elements. What is the chemical formula for a molecule of each of these elements?*

for the element oxygen is O. But in nature, oxygen occurs as a molecule that contains 2 atoms of oxygen bonded together. So the formula for a molecule of oxygen is O_2. Some other gases that exist only in pairs of atoms are hydrogen, H_2, nitrogen, N_2, fluorine, F_2, and chlorine, Cl_2. Remember that the symbols for the elements just listed are the letters only. The formulas are the letters with the small number 2 at the lower right.

When writing a chemical formula, you use the symbol of each element in the compound. You also use small numbers, called **subscripts.** Subscripts are placed to the lower right of the symbols. A subscript gives the number of atoms of the element in the compound. When there is only 1 atom of an element, the subscript 1 is not written. It is understood to be 1.

Carbon dioxide is a compound of the elements carbon and oxygen. Its formula is CO_2. By looking at the formula, you can tell that every molecule is made up of 1 atom of carbon (C) and 2 atoms of oxygen (O). Sulfuric acid has the formula H_2SO_4. How many hydrogen atoms, sulfur atoms, and oxygen atoms are there in a molecule of sulfuric acid?

Can you now see the advantages of using chemical formulas? Not only does a formula save space, but it tells you a lot about the compound. It tells you the elements that make up the compound. And it tells you how many atoms of each element combine to form the compound.

Chemical Equations

If you think of chemical symbols as "letters" and chemical formulas as "words," then you can write chemical "sentences." Chemical sentences are a way to describe a chemical process, or chemical reaction. As you learned in Chapter 2, during a chemical reaction, substances are changed into new and different substances through a rearrangement of their atoms. New chemical substances with new properties

ACTIVITY

CALCULATING

Count the Atoms

List the elements that are present in each of the following compounds. Calculate how many atoms of each element are present in each compound.

NaHCO$_3$
C$_2$H$_4$O$_2$
Mg(OH)$_2$
3H$_3$PO$_4$

are formed. By using chemical symbols and formulas, you can describe a chemical reaction.

Have you ever seen charcoal burning in a barbecue grill? If so, you were watching a chemical reaction. The carbon atoms in the charcoal were combining with the oxygen molecules in the air to form the gas carbon dioxide. The reaction could be written:

carbon atoms plus oxygen molecules
produce carbon dioxide molecules

By using symbols and formulas, this reaction can be written in a simpler way:

$$C + O_2 \longrightarrow CO_2$$

The symbol C represents an atom of carbon. The formula O_2 represents a molecule of oxygen. And the formula CO_2 represents a molecule of carbon dioxide. The arrow is read "yields," which is another way of saying "produces."

The description of a chemical reaction using symbols and formulas is called a **chemical equation.** An equation is another example of chemical shorthand. Instead of using words to describe a chemical reaction, you can use a chemical equation.

Here is another example. The chemical equation for the formation of water from the elements hydrogen and oxygen is

$$H_2 + O_2 \longrightarrow H_2O$$

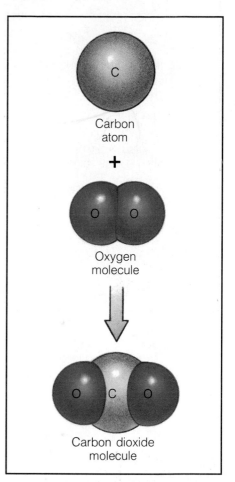

Figure 3–21 *This illustration shows the chemical reaction that occurs when carbon and oxygen combine to form carbon dioxide. What is the chemical formula for carbon dioxide?*

PROPERTIES OF ELEMENTS, COMPOUNDS, AND MIXTURES

Elements	Compounds	Mixtures
Made up of only one kind of atom	Made up of more than one kind of atom	Made up of more than one kind of molecule
Cannot be broken down by chemical means	Can be broken down by chemical means	Can be separated by physical means
Has same properties as atoms making it up	Has different properties from elements making it up	Has same properties as substances making it up
Has same properties throughout	Has same properties throughout	Has different properties throughout

Figure 3–22 *This table shows the common properties of elements, compounds, and mixtures. Which of the three substances does not have the same properties throughout?*

Figure 3–23 *During the formation of water, 2 hydrogen molecules combine with a molecule of oxygen to form 2 water molecules. What is the chemical equation for this reaction?*

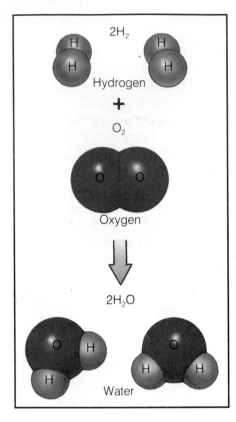

2H₂

Hydrogen

+

O₂

Oxygen

2H₂O

Water

This equation ($H_2 + O_2 \longrightarrow H_2O$) tells you what elements are combining and what product is formed. But something is wrong with this equation. Do you know what it is?

Look at the number of oxygen atoms on each side of the equation. Are they equal? On the left side of the equation there are 2 oxygen atoms. On the right side there is only 1 oxygen atom. Could 1 oxygen atom have disappeared? Scientists know that atoms are never created or destroyed in a chemical reaction. Atoms can only be rearranged. So there must be the same number of atoms of each element on each side of the equation. The equation must be balanced. The balanced equation for the formation of water is

$$2H_2 + O_2 \longrightarrow 2H_2O$$

Now count the atoms of each element on each side of the equation. You will find that they are the same: 4 atoms of hydrogen on the left side and on the right; 2 atoms of oxygen on the left side and on the right. The equation is correctly balanced.

You have seen that an equation can be balanced by placing the appropriate number in front of a chemical formula in the equation. This number is called a **coefficient** (koh-uh-FIHSH-uhnt). The correctly balanced equation now tells you that 2 molecules of hydrogen combine with 1 molecule of oxygen to produce 2 molecules of water. A balanced equation is evidence of a chemical reaction.

3–4 Section Review

1. What is a compound?
2. How is a compound different from an element?
3. What is a molecule? How is a molecule of an element or a compound represented?
4. Why must a chemical equation be balanced?

Connection—*You and Your World*
5. How is a recipe for baking a cake like a chemical equation? What three things must be included in a good, easy-to-follow recipe?

CONNECTIONS

Time and a Tree

You have to hurry. . . . In three minutes you will be late for class. Time has a way of ruling your days. Time tells you where to be, and when. Time also tells you what to do. Days seem short when you have a lot to do.

Time for a sugar maple is different. You probably do not give much thought to time's effects on a tree. A tree stands tall and proud for so long that time seems to have little impact on its life from one day to the next. Yet if you were able to observe a tree over a year's time, you would notice profound changes.

Take this sugar maple in the photograph. It is the middle of March and the Vermont woods where this tree grows are covered with a winter's blanket of snow. Few birds can be seen or heard, and even fewer other animals are in sight. For most forms of forest life, winter is a time of rest. But deep within the trunk of a maple tree, things are beginning to stir. Sap, a sweet juice conducted in tiny tubes in the tree trunk, is starting to move upward from the roots. Farmers are able to "tap" the tree's trunk and collect buckets of sap. Later the sap will be boiled down into delicious maple syrup. But where did this sweet sap come from?

Actually, the sugars in the sap were made by the tree the year before. In the previous spring, the maple tree leafed out. The beautiful leaves that clothed the maple then are efficient collectors of the sun's energy. The green pigment in the leaves, which is called chlorophyll, trapped the energy of sunlight and used it to make food. In a complex series of chemical reactions, water was combined with carbon dioxide to produce sugars. Scientists write a chemical equation to describe the process:

$$6CO_2 + 6H_2O \xrightarrow[\text{(chlorophyll)}]{\text{(sunlight)}} C_6H_{12}O_6 + 6O_2$$

Carbon dioxide plus water plus the energy of the sun trapped by chlorophyll yields sugar plus oxygen—a simple equation that describes the single most important chemical reaction on Earth.

You might wonder why this reaction is so important. It is important because plants are able to use the energy of the sun in ways animals never can. And in so doing, plants produce food (in the form of plant sugars) and oxygen necessary for the survival of animals on Earth.

Laboratory Investigation

Making Models of Chemical Reactions

Problem

How do atoms and molecules of elements and compounds combine in chemical reactions?

Materials *(per group)*

> toothpicks
> red, yellow, green, blue, purple (red-blue), and orange (yellow-red) food coloring
> 25 large marshmallows

Procedure

A. *Making Marshmallow Atoms*

1. Prepare model atoms by applying food coloring to the marshmallows as follows:

   N (nitrogen)—red (2)

   H (hydrogen)—blue (6)

   Cu (copper)—green (4)

   O (oxygen)—yellow (7)

   K (potassium)—orange (2)

   Cl (chlorine)—purple (2)

2. Let the marshmallows dry for 2 hours.

B. *Assembling the Marshmallow Molecules*

1. Use a toothpick to join two red marshmallows to make a molecule of N_2. Use a toothpick to join two blue marshmallows to make a molecule of H_2.

2. Ammonia (NH_3) is used in cleaning solutions and in the manufacture of fertilizers. A molecule of ammonia contains 1 nitrogen atom and 3 hydrogen atoms. Use the marshmallow molecules of nitrogen and hydrogen you made in step 1 to form an ammonia molecule. You may use as many nitrogen and hydrogen molecules as you need to make ammonia molecules as long as you do not have any atoms left over. Remember, hydrogen and nitrogen must start out as molecules consisting of 2

atoms each. Now balance the equation for the chemical reaction that produces ammonia:

$$\underline{\quad}N_2 + \underline{\quad}H_2 \longrightarrow \underline{\quad\quad}NH_3$$

3. Use two green marshmallows for copper and one yellow marshmallow for oxygen to make a model of a copper oxide molecule (Cu_2O). With a white marshmallow representing carbon, manipulate the marshmallow models to illustrate the reaction below, which produces metallic copper. Balance the equation.

$$\underline{\quad}Cu_2O + \underline{\quad}C \longrightarrow \underline{\quad}Cu + \underline{\quad}CO_2$$

4. Use orange for potassium, purple for chlorine, and yellow for oxygen to assemble a molecule of potassium chlorate ($KClO_3$).

Observations

1. How many molecules of N_2 and H_2 are needed to produce 2 molecules of NH_3?

2. How many molecules of copper are produced from 2 molecules of Cu_2O?

Analysis and Conclusions

1. Which substances that you made are elements? Which are compounds?

2. If you had to make 5 molecules of ammonia (NH_3), how many red marshmallows would you need? How many blue marshmallows?

Study Guide

Summarizing Key Concepts

3–1 Classes of Matter

▲ Matter is classified according to its makeup as mixtures, solutions, elements, or compounds.

3–2 Mixtures

▲ A mixture is composed of two or more substances mixed together but not chemically combined.

▲ The substances in a mixture can be present in any amount.

▲ The substances in a mixture can be separated by simple physical means.

▲ A mixture that does not appear to be the same throughout is a heterogeneous mixture.

▲ A mixture that appears to be the same throughout is a homogeneous mixture.

▲ The particles in a colloid, a type of homogeneous mixture, are not dissolved.

▲ A solution is a type of homogeneous mixture formed when one substance, called the solute, dissolves in another substance, called the solvent.

▲ The amount of a solute that can completely dissolve in a given solvent at a specific temperature is called its solubility.

▲ Alloys are metal solutions in which solids are dissolved in solids.

3–3 Elements

▲ A pure substance is made of only one kind of material, has definite properties, and is the same throughout.

▲ Elements are the simplest type of pure substance. They cannot be broken down into simpler substances without losing their identity.

▲ Elements are made of atoms, which are the building blocks of matter.

3–4 Compounds

▲ Compounds are two or more elements chemically combined.

▲ Most compounds are made of molecules. A molecule is made of two or more atoms chemically bonded together.

▲ A chemical formula, which is a combination of chemical symbols, usually represents a molecule of a compound.

▲ A chemical equation describes a chemical reaction.

Reviewing Key Terms

Define each term in a complete sentence.

3–2 Mixtures
mixture
heterogeneous mixture
homogeneous mixture
colloid
solution
solute
solvent
soluble
insoluble
solubility
alloy

3–3 Elements
pure substance
element
atom
chemical symbol

3–4 Compounds
compound
molecule
chemical formula
subscript
chemical equation
coefficient

Chapter Review

Content Review

Multiple Choice

Choose the letter of the answer that best completes each statement.

1. Matter that consists of two or more substances mixed together but not chemically combined is called a(an)
   a. element.   c. pure substance.
   b. compound.   d. mixture.
2. An example of a heterogeneous mixture is
   a. salt water.   c. stainless steel.
   b. salad dressing. d. salt.
3. In a solution, the substance being dissolved is called the
   a. solvent.   c. solubility.
   b. solute.   d. insoluble.
4. The simplest type of pure substance is a (an)
   a. compound.   c. solution
   b. alloy.   d. element.

5. The chemical formula for a molecule of nitrogen is
   a. N.   c. N_3.
   b. N_2.   d. Ni.
6. Pure substances that are made of more than one element are called
   a. compounds.   c. alloys.
   b. mixtures.   d. solutions.
7. Which of the following is not an alloy?
   a. zinc   c. stainless steel
   b. gold jewelry   d. brass
8. Which of the following is not a compound?
   a. H_2 c. H_2SO_4
   b. H_2O d. CO_2

True or False

If the statement is true, write "true." If it is false, change the underlined word or words to make the statement true.

1. The basic building block of matter is the <u>compound</u>.
2. When elements combine to form compounds, their properties <u>do not</u> change.
3. One example of a <u>homogeneous</u> mixture is concrete.
4. Substances in a <u>mixture</u> keep their separate identities and most of their own properties.
5. Mixtures can be separated by simple <u>chemical</u> means.
6. Salt and sugar are <u>insoluble</u> in water.
7. The "best mixed" of mixtures is a <u>solution</u>.
8. The particles in a <u>colloid</u> are mixed together but not dissolved.

Concept Mapping

Complete the following concept map for Section 3–1. Refer to pages N6–N7 to construct a concept map for the entire chapter.

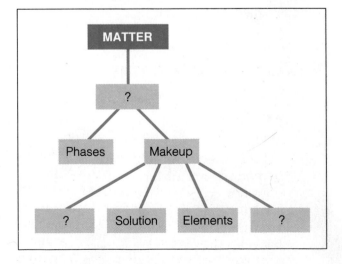

Concept Mastery

Discuss each of the following in a brief paragraph.

1. A solution is classified as a mixture instead of a compound. Why?
2. Describe a method you can use to separate the following mixtures. Your method should take into account the physical properties of the substances that make up the mixture.
   a. sugar and water
   b. powdered iron and powdered aluminum
   c. wood and gold
   d. nickels and dimes
3. What two things does a chemical formula indicate about a compound?
4. Write the symbols for the following elements and describe one use for each: (a) gold, (b) oxygen, (c) carbon, (d) hydrogen, (e) silver, (f) calcium, (g) nitrogen.
5. Chemical equations can be balanced by using coefficients but not subscripts. Explain why this is so.
6. What is a solution? What two parts make up a solution? Describe three properties of a solution.

Critical Thinking and Problem Solving

Use the skills you have developed in this chapter to answer each of the following.

1. **Making inferences** Suppose you find a chemistry book that is written in a language you cannot understand. When you look at the book, however, you realize that you can understand and duplicate the chemical reactions printed in the text. Explain how you can do this?
2. **Making calculations** Balance the following equations:
   a. $Mg + O_2 \longrightarrow MgO$
   b. $NaCl \longrightarrow Na + Cl_2$
   c. $CH_4 + O_2 \longrightarrow CO_2 + H_2O$
   d. $H_2 + O_2 \longrightarrow H_2O$
3. **Designing an experiment** Describe an experiment to show that
   a. water is a compound, not an element.
   b. salt water is a solution, not a pure substance.
4. **Relating facts** You learned that mixtures have three important properties. Use the example of a bowl of cereal with milk and strawberries to illustrate each property.
5. **Classifying data** Develop a classification system for the months of the year. State the property or properties that you will use to classify the months. Do not use the four seasons. Try to make your system as useful and as specific as possible.
6. **Making inferences** Explain whether or not you believe there exists a true "universal solvent" capable of dissolving all other substances. Include a description of the kind of container you would need to hold such a solvent.
7. **Using the writing process** Water has been accused of being an element by an assistant district attorney. You are the defense attorney whose job is to convince the jury that water is a compound. Write up a brief summation to the jury to make your case. Use only the very best scientific information in your case.

Atoms: Building Blocks of Matter

Beads of mercury gleam on a sheet of cloth. Some beads are small; others are large. But they are all still mercury, a pure substance. If you were to take the smallest bead of mercury and slice it in half once, twice, three times, even a thousand times, you would still be left with mercury—or would you?

Is there some incredibly tiny bead of mercury that if sliced one more time would no longer be mercury? It was just this kind of question that sparked the imagination and curiosity of early scientists. They hypothesized and argued as the years passed—for more than 2000 years, in fact.

Then, slowly, clues were found. Experiments were performed. New ideas were explored. And finally an answer was developed. In many ways it was quite a simple answer. To find it out, let's begin at the beginning.

Journal *Activity*

You and Your World Have you ever seen pictures of the Great Pyramids that were built in Egypt thousands of years ago? Block upon block, these great tombs were constructed with few tools and great effort. Even the tallest building is made of small parts. Draw a picture in your journal of a building you like. Next to your drawing, make another drawing of one of the pieces that make up your building.

◀ *Silvery beads of liquid mercury gleam in the photographer's lights.*

4–1 An Atomic Model of Matter

In the last few chapters, you learned several important facts about matter. All materials are made of matter. Matter is anything that has mass and volume. And the basic building blocks of matter are atoms. You have learned all of this in a rather short time. But the story of how scientists have come to know what they do about matter spans a much greater time period—thousands of years to be sure!

For more than 2400 years, philosophers and scientists have tried to determine the composition of matter using a variety of experiments and observations. Because the basic building blocks of matter (atoms) could not until recently be seen, researchers have relied on observations of how matter behaves. Such observations are called indirect evidence.

Indirect evidence about an object is evidence you get without actually seeing or touching the object. As you gather indirect evidence, you can develop a mental picture, or model. A model uses familiar ideas to explain unfamiliar facts observed in nature. A model can be changed as new information is collected. As you read further, you will learn how a model of matter was developed and changed over many years. From the early Greek concept of the atom to the modern atomic theory, scientists have built on and modified existing models of the atom. Let's see just how.

Figure 4–1 *Scientists often depend on indirect evidence to develop a model of something that cannot be observed directly. Use the two drawings in this figure to develop a model that might explain what happened during the few hours that separate the two drawings.*

Figure 4–2 *Unlike many artists working today, ancient Greek artists did not paint on canvas. This vase painting, showing a warrior carrying the body of a dead companion, was made about the time Democritus lived.*

The Greek Model

The search for a description of matter began with the Greek philosopher Democritus (dih-MAHK-ruh-tuhs) more than 2000 years ago. He and many other philosophers had puzzled over this question: Could matter be divided into smaller and smaller pieces forever, or was there a limit to the number of times a piece of matter could be divided?

After much observation and questioning, Democritus concluded that matter could not be divided into smaller and smaller pieces forever. Eventually the smallest possible piece would be obtained. This piece would be indivisible. Democritus named this smallest piece of matter an atom. The word *atom* comes from the Greek word *atomos*, meaning "not to be cut," or "indivisible."

The Greek philosophers who shared Democritus' belief about the atom were called atomists. The atomists had no way of knowing what atoms were or how they looked. But they hypothesized that atoms were small, hard particles that were all made of the same material but were of different shapes and sizes. Also, they hypothesized that they were infinite in number, always moving, and capable of joining together.

Although Democritus and the other atomists were on the right trail, the theory of atoms was ignored and forgotten. Few people believed the idea. In fact, it took almost 2100 years before an atomic model of matter was accepted.

Shine a Little Light on It

Indirect evidence about an object is evidence you get without actually seeing or touching the object. As you gather indirect evidence, you can develop a model, or mental picture.

1. Fill two glasses almost completely full with water. Leave one glass as is. Add a piece of soap about the size of a pea to the other glass. Stir the water to dissolve the soap.

2. Turn off the lights in the room. Make sure that the room is completely dark.

3. Shine a flashlight beam horizontally from the side of the glass into the soapy water. Aim the light beam so that it enters the water just below the surface. Repeat this procedure with the glass of plain water. Observe and record the effect of the light beam in each glass.

What was the effect in the plain water? What was the effect in the soapy water? What caused the effect in the soapy water? What was the role of the glass of plain water in this activity?

Figure 4–3 *The Rosetta Stone was discovered in 1799, a few years before Dalton's atomic theory was proposed. In his work, Dalton tried to explain the mysteries of atomic structure. Do you know how the Rosetta Stone helped to explain the mysteries of Egyptian writing?*

Dalton's Model

In the early 1800s, the English chemist John Dalton performed a number of experiments that eventually led to the acceptance of the idea of atoms. Dalton had long been interested in meteorology, the study of weather. His observations about the composition of air led him to investigate the properties of gases. He discovered that gases combine as if they were made of individual particles. These particles were the atoms of Democritus.

In 1803, Dalton combined the results of his experiments with other observations about matter and proposed an atomic theory. The basic ideas of Dalton's atomic theory are as follows:

- **All elements are composed of atoms. Atoms are indivisible and indestructible particles.**
- **Atoms of the same element are exactly alike.**
- **Atoms of different elements are different.**
- **Compounds are formed by the joining of atoms of two or more elements.**

Dalton's atomic theory of matter became one of the foundations of modern chemistry. But like many scientific theories, Dalton's theory had to be modified as scientists gained more information about the structure of matter.

Figure 4–4 *The atoms of relatively few elements make up everything in the world around you—whether it be a tranquil country landscape or a vibrant (but smoggy) city.*

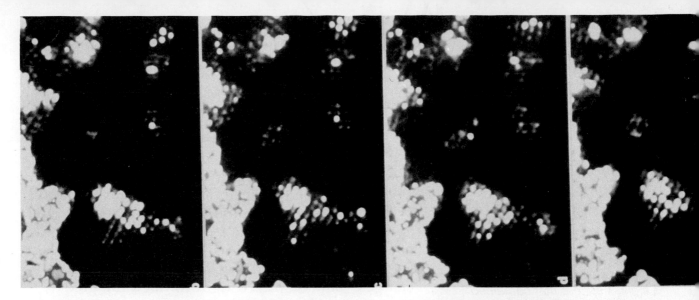

Thomson's Model

Was Dalton's theory correct? Is an atom indivisible? In 1897, the English scientist J. J. Thomson provided the first hint that an atom is made of even smaller particles. Thomson was studying the passage of an electric current through a gas. The gas gave off rays that Thomson showed were made of negatively charged particles. But the gas was known to be made of uncharged atoms. So where had the negatively charged particles come from? From within the atom, Thomson reasoned. A particle smaller than the atom had to exist. The atom was divisible! Thomson called the negatively charged particles "corpuscles." Today these particles are known as electrons.

As often happens in science, Thomson's discovery of electrons created a new problem to solve. The atom as a whole was known to be uncharged, or neutral. But if electrons in the atom were negatively charged, what balanced the negative charge? Thomson reasoned that the atom must also contain positively charged particles. But try as he might, he was unable to find this positive charge.

Thomson was so certain that these positively charged particles existed that he proposed a model of the atom that is sometimes called the "plum pudding" model. Figure 4–6 shows Thomson's model. **According to Thomson's atomic model, the atom was made of a puddinglike positively charged material throughout which negatively charged electrons were scattered, like plums in a pudding.** Thomson's

Figure 4–5 *These photographs of uranium atoms were taken by scientists working at the University of Chicago. The blue, yellow, and red spots are uranium atoms magnified more than 5 million times. In this remarkable series of photographs, you can observe the actual movement of the atoms.*

Figure 4–6 *Thomson's model of the atom pictured a "pudding" of positively charged material. Negatively charged electrons were scattered throughout like plums in a pudding. What is the overall charge on this atom? Why?*

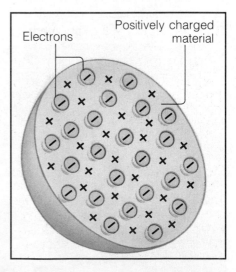

Electrons

Positively charged material

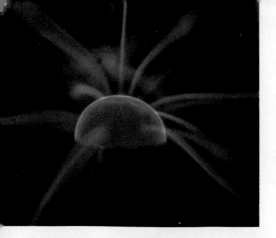

Figure 4–7 *Artist and physicist Bill Parker created this "electric art" by passing an electric current through a glass sphere that contained certain gases. The light is produced when electrons in the gases absorb energy and release it. Who is credited with the discovery of the electron?*

Figure 4–8 *In Rutherford's experiment, most of the positively charged material passed right through the gold sheet (left). A few particles were slightly deflected, and even fewer particles bounced straight back. From these observations, Rutherford concluded that the atom was mostly empty space with a dense positively charged nucleus in the center (right).*

model, while far from correct, was an important step in understanding the structure of the atom.

Rutherford's Model

In 1908, the English physicist Ernest Rutherford was hard at work on an experiment that seemed to have little to do with unraveling the mysteries of atomic structure. Rutherford's experiment involved firing a stream of tiny positively charged particles at a thin sheet of gold foil. (Although the gold foil was hammered very thin, it was still 2000 atoms thick!) Rutherford discovered that most of the positively charged "bullets" passed right through the gold atoms in the sheet of foil without changing course at all. This could only mean that the gold atoms in the sheet were mostly empty space! Atoms were not a pudding filled with a positively charged material, as Thomson had thought.

Some of the "bullets," however, did bounce away from the gold sheet as if they had hit something solid. In fact, a few bounced almost straight back. What could this mean? Rutherford knew that positive charges repel other positive charges. So he proposed that an atom had a small, dense, positively charged center that repelled his positively charged "bullets." He called this center of the atom the **nucleus** (NOO-klee-uhs; plural: nuclei, NOO-klee-igh). The nucleus is tiny compared to the atom as a whole. To get an idea of the size of the nucleus in an atom, think of a marble in a baseball stadium!

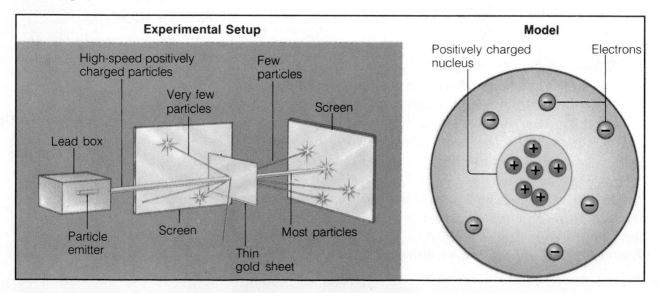

Rutherford reasoned that all of an atom's positively charged particles were contained in the nucleus. The negatively charged electrons were scattered outside the nucleus around the atom's edge. Between the nucleus and the electrons was mostly empty space! Although this model was useful in many ways, it did not adequately explain the arrangement of the electrons. It would be the job of future scientists to improve on the Rutherford atomic model.

The Bohr Model

Rutherford's model proposed that negatively charged electrons were held in an atom by the attraction between them and the positively charged nucleus. But where exactly were the electrons in the atom? In 1913, the Danish scientist Niels Bohr proposed an improvement on the Rutherford model. In his model he placed each electron in a specific energy level. **According to Bohr's atomic model, electrons move in definite orbits around the nucleus, much like planets circle the sun. These orbits, or energy levels, are located at certain distances from the nucleus.**

The Wave Model

Bohr's model worked well in explaining the structure and behavior of simple atoms such as hydrogen. But it did not explain more complex atoms.

Today's atomic model is based on the principles of wave mechanics. The basic ideas of wave mechanics are complicated and involve complex mathematical equations. Some of the conclusions of this theory, however, will help you understand the arrangement of electrons in an atom.

According to the theory of wave mechanics, electrons do not move about an atom in a definite path

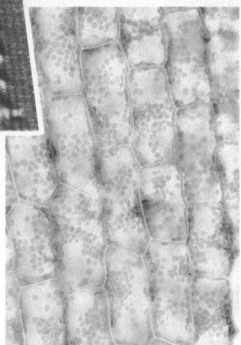

Figure 4–9 *Atoms are the incredibly small building blocks of matter. This photograph shows the first structure ever built atom by atom (left). Cells are the building blocks of all living things (right). But even the largest cells are made of tiny atoms.*

ACTIVITY
DOING

Making a Model Atom

1. Use materials such as cardboard, construction paper, colored pencils, string, and cotton to construct models of the Thomson atom and the Rutherford atom.

2. Label the models and place them on display in your classroom. Write a brief description of the experiment that each model was based on.

Figure 4–10 *This atomic model shows the nucleus with its neutrons and protons. Surrounding the nucleus are rapidly moving electrons. Can scientists know with certainty where a particular electron is located in an atom?*

like planets around the sun. In fact, it is impossible to determine the exact location of an electron. Scientists can only predict where an electron is most likely to be found. The probable location of an electron is based on how much energy the electron has.

As you can see, the modern atomic model is based on the models of Rutherford and Bohr, and on the principles of wave mechanics. **According to the modern atomic model, an atom has a small positively charged nucleus surrounded by a large region in which there are enough electrons to make the atom neutral.**

Figure 4–11 *When atomic particles collide, new and unusual particles may be produced. By studying the tracks made by these particles in a bubble chamber, scientists can learn more about the nature and interactions of atomic particles.*

4–1 Section Review

1. How has the model of the atom changed over time?
2. Why is indirect evidence important in studying the structure of the atom?
3. What atomic particle did J. J. Thomson discover?
4. What is the center of the atom called? How was it discovered?
5. How does the wave model of electron placement differ from the model of electron position proposed by Niels Bohr?

Connection—*Science and Technology*
6. The model that explains atomic structure has changed over time. How has technology contributed to this change?

CONNECTIONS

Data In . . . A Building Out

You now know that as a result of many, many years of research, scientists have developed a model that describes the structure of the atom—and they have done so without ever having seen an actual atom. Today, scientists have abundant evidence that this model of the atom is accurate.

But making a model has other, more practical uses. You can see examples of these uses almost daily as you walk down streets in your town or city. Today, models developed by *computers* serve as blueprints for the houses and buildings you see around you. Such a model can show how a building will look while it is still only a series of ideas in an architect's mind and a few quick sketches on a sheet of paper.

Computer programs offer architects and building engineers a wide range of applications: designing a building, determining air flow and people movement, measuring wind effect on structure, to name a few. Some computer programs can even show the effects of light and shadow that will exist in a finished building. If such effects are undesirable, changes in the location and number of windows and doors, for example, can be made long before construction of the building begins.

Remember, however, that computer-assisted designs are only a tool to make the work of architects and engineers easier. The design of a wonderful building—a building that stirs the heart and lifts the spirit—still begins deep within the human mind and results from the combined talents of its creators.

Computers have become invaluable tools in many different professions. The final designs and plans for this building were developed with the help of a computer.

4–2 Structure of the Atom

When Thomson performed his experiments, he was hoping to find a single particle smaller than an atom. This task is similar to finding a particular grain of sand among the grains of sand making up all the beaches of the Earth! Certainly Thomson would be surprised to learn that today about 200 different kinds of such particles are known to exist! Because these particles are smaller than an atom, they are called **subatomic particles.**

The three main subatomic particles are the proton, the neutron, and the electron. As you read about these particles, note the location, mass, and charge of each. In this way, you will better understand the modern atomic theory. Let's begin with the nucleus, or center, of the atom.

The Nucleus

The nucleus is the "core" of the atom, the center in which 99.9 percent of the mass of the atom is located. Yet the nucleus is about a hundred thousand times smaller than the entire atom! In fact, the size of the nucleus compared to the entire atom has been likened to the size of a bee compared to a football stadium! Two of the three main subatomic particles are found in the nucleus.

PROTONS Those positively charged "bullets" that Rutherford fired at the gold sheet bounced back because of **protons** in the nucleus of the gold atoms. Protons are positively charged particles found in the nucleus. All protons are identical, regardless of the element in which they are found.

Figure 4–12 *A lithium nucleus contains 3 protons and 4 neutrons. A carbon nucleus contains 6 protons and 6 neutrons. How many electrons does a lithium nucleus contain? A carbon nucleus?*

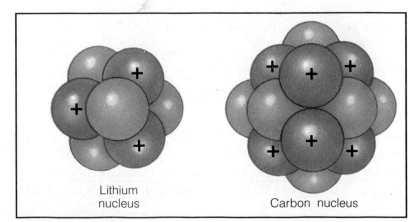

Lithium nucleus

Carbon nucleus

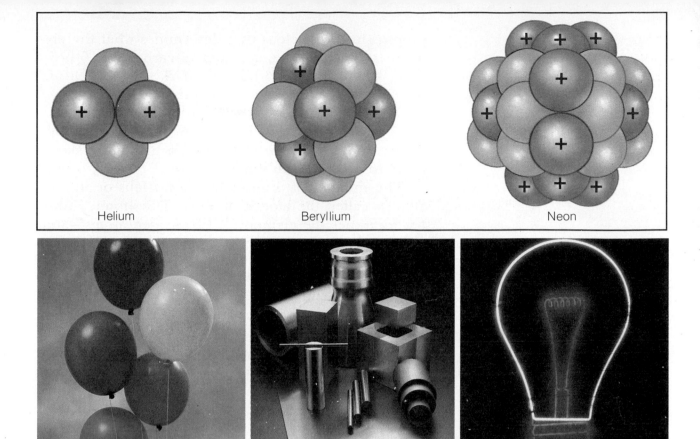

Helium

Beryllium

Neon

Because the masses of subatomic particles are so small, scientists use a special unit to measure them. They call this unit an **atomic mass unit,** or amu. The mass of a proton is 1 amu. To get a better idea of how small a proton is, imagine the number 6 followed by 23 zeros. It would take that many protons to equal a mass of just 1 gram!

NEUTRONS Sharing the nucleus with the protons are the electrically neutral **neutrons.** Neutrons have no charge. Like protons, all neutrons are identical. Neutrons have slightly more mass than protons. But the mass of a neutron is still considered to be 1 amu.

Figure 4–13 *The nuclei of helium, beryllium, and neon atoms all contain protons and neutrons. Yet helium, beryllium, and neon are very different elements. What accounts for these differences?*

Atomic Number

You read before that atoms of different elements are different. But if all protons are identical and all neutrons are identical, what accounts for these differences? The answer lies in the particles found in the nucleus—more specifically, in the number of protons in the nucleus. Because it is the number of protons in the nucleus that determines what the element is. For example, an atom of carbon has 6 protons in its nucleus. Carbon is a dark solid. Much of the sooty remains of a burned piece of wood are made up of atoms of carbon. An atom of nitrogen has 7 protons in its nucleus—only one more proton than carbon. Nitrogen is a colorless gas that makes up most of the Earth's atmosphere.

The number of protons in the nucleus of an atom is called the **atomic number.** The atomic number identifies the element. All hydrogen atoms—and only hydrogen atoms—have 1 proton and an atomic number of 1. Helium atoms have an atomic number of 2. There are 2 protons in the nucleus of every helium atom. Oxygen has an atomic number of 8, and 8 protons are in the nucleus of each atom. How many protons does an atom of uranium—atomic number 92—have? And what are the atomic numbers of the elements carbon and nitrogen that you just read about?

Isotopes

The atomic number of an element will never change. This means that there is always the same number of protons in the nucleus of every atom of that element. But the number of neutrons is not so constant! Atoms of the same element can have different numbers of neutrons.

Atoms of the same element that have the same number of protons but different numbers of neutrons are called **isotopes** (IGH-suh-tohps). Look at Figure 4–15. You will see three different isotopes of

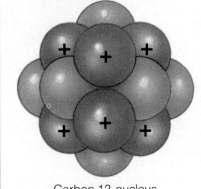

Carbon-12 nucleus

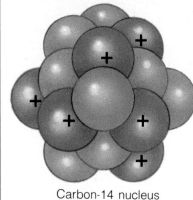

Carbon-14 nucleus

Figure 4–14 *These two isotopes of carbon have the same atomic number, 6. Although it cannot speak, this fossil fish can be made to divulge its age. To do this, scientists analyze the proportions of carbon isotopes present in the fish. What is the difference between the two carbon isotopes?*

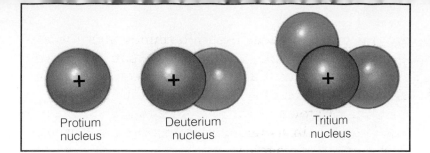

Protium
nucleus

Deuterium
nucleus

Tritium
nucleus

Figure 4–15 *The three isotopes of hydrogen are protium, deuterium, and tritium. Which isotope contains 2 neutrons? What is the atomic number of each isotope?*

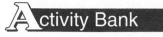

Activity Bank

Hunting for Treasure in Trash, p.158

the element hydrogen. Note that the number of protons does not change. Remember that the atomic number, or number of protons, identifies a substance. No matter how many neutrons there are in the nucleus, 1 proton always means the atom is hydrogen. How many neutrons does each hydrogen isotope have?

Mass Number and Atomic Mass

All atoms have a **mass number.** The mass number of an atom is the sum of the protons and neutrons in its nucleus. The mass number of the carbon isotope with 6 neutrons is 6 (protons) + 6 (neutrons), or 12. The mass number of the carbon isotope with 8 neutrons is 6 (protons) + 8 (neutrons), or 14. To distinguish one isotope from another, the mass number is given with the element's name.

Two common isotopes of the element uranium are uranium-235 and uranium-238. The atomic number—or number of protons—of uranium is 92. Since the mass number is equal to the number of protons plus the number of neutrons, the number of neutrons can easily be determined. The number of neutrons is determined by subtracting the atomic number (number of protons) from the mass number (number of protons + neutrons). Here are two problems for you to try. How many neutrons are there in uranium-235? In uranium-238?

Any sample of an element as it occurs in nature contains a mixture of isotopes. As a result, the **atomic mass** of the element is the average mass of all the isotopes of that element as they occur in nature. For this reason, the atomic mass of an element is not usually a whole number. For example, the atomic mass of carbon is 12.011. This number indicates that there are more atoms of carbon-12 than there are of carbon-14. Can you explain why this conclusion is reasonable?

Figure 4–16 *This chart shows the symbol, atomic number, and mass number for some common elements. Why is the mass number always a whole number but the atomic mass is not?*

COMMON ELEMENTS		
Name	Atomic Number	Mass Number
Hydrogen H	1	1
Helium He	2	4
Carbon C	6	12
Nitrogen N	7	14
Oxygen O	8	16
Fluorine F	9	19
Sodium Na	11	23
Aluminum Al	13	27
Sulfur S	16	32
Chlorine Cl	17	35
Calcium Ca	20	40
Iron Fe	26	56
Copper Cu	29	64
Zinc Zn	30	65
Silver Ag	47	108
Gold Au	79	197
Mercury Hg	80	201
Lead Pb	82	207

PROBLEM Solving

Improving the Odds

You are trying to locate a friend on a sunny Saturday afternoon. Although you cannot say with absolute certainty where your friend is, you can estimate the chance of finding your friend in various places. Your estimates are based on past experiences.

Create a chart that lists at least seven possible locations for your friend. Next to each location, give the probability of finding your friend there. Express the probability in percent. For example, there is a 50-percent probability that your friend is at the school yard playing soccer. Remember that the total probability for the seven events should equal 100 percent.

Would a change in the weather influence your probability determination? How about a change in the day of the week? In the season of the year?

Relating Concepts

How does this activity help you better understand the concept of probability? How does it relate to an electron's location in an atom?

The Electrons

If you think protons and neutrons are small, picture this. Whirling around outside the nucleus are particles that are about 1/2000 the mass of either a proton or a neutron! These particles are **electrons.** Electrons have a negative charge and a mass of 1/1836 amu. In an uncharged atom, the number of negatively charged electrons is equal to the number of positively charged protons. The total charge of the atom is zero. Thus the atom is said to be neutral.

As you have learned, electrons do not move in fixed paths around the nucleus. In fact, the exact location of an electron cannot be known. Only the probability, or likelihood, of finding an electron at a particular place in an atom can be determined.

In fact, the entire space that the electrons occupy is what scientists think of as the atom itself. Sometimes this space is called the **electron cloud.** But do not think of an atom as a solid center surrounded by a fuzzy, blurry cloud. For the electron cloud is a space in which electrons are likely to be found. It is somewhat like the area around a beehive in which the bees move. Sometimes the electrons are near the nucleus; sometimes they are farther away from it. In a hydrogen atom, 1 electron "fills" the cloud. It fills the cloud in the sense that it can be found almost anywhere within the space.

Although the electrons whirl about the nucleus billions of times in one second, they do not do so in a random way. Each electron seems to be locked into a certain area in the electron cloud. The location of an electron in the cloud depends upon how much energy the electron has.

According to modern atomic theory, electrons are arranged in **energy levels.** An energy level represents the most likely location in the electron cloud in which an electron can be found. Electrons with the lowest energy are found in the energy level closest to the nucleus. Electrons with higher energy are found in energy levels farther from the nucleus.

ACTIVITY

CALCULATING

The Mystery Element

You can identify the mystery element by performing the following mathematical calculations.

a. Multiply the atomic number of hydrogen by the number of electrons in mercury, atomic number 80.

b. Divide this number by the number of neutrons in helium, atomic number 2, mass number 4.

c. Add the number of protons in potassium, atomic number 19.

d. Add the mass number of the most common isotope of carbon.

e. Subtract the number of neutrons in sulfur, atomic number 16, mass number 32.

f. Divide by the number of electrons in boron, atomic number 5, mass number 11.

Which of the following elements is the mystery element: fluorine, atomic number 9; neon, atomic number 10; or sodium, atomic number 11?

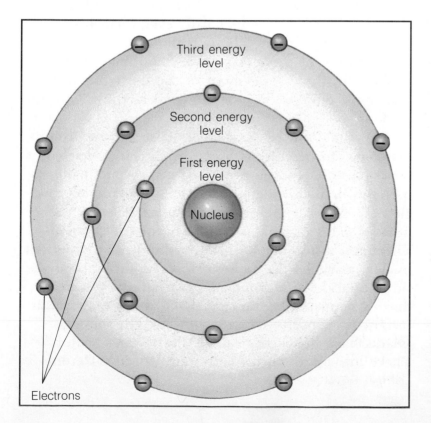

Electrons

Figure 4–17 *Each energy level in an atom can hold only a certain number of electrons. How many electrons are in the first, second, and third energy levels shown here?*

SUBATOMIC PARTICLES			
Particle	Mass (amu)	Charge	Location
Proton	1	+	Nucleus
Neutron	1	Neutral	Nucleus
Electron	$1/1836$	–	Electron cloud

Figure 4–18 *The mass, charge, and location of the three basic subatomic particles are shown in this chart. Which subatomic particle has a neutral charge and a mass of 1 amu? Where is this subatomic particle located?*

Each energy level within an atom can hold only a certain number of electrons. The energy level closest to the nucleus—the lowest energy level—can hold no more than 2 electrons. The second energy level can hold 8 electrons. The third energy level can hold 18 electrons.

The properties of the different elements depend upon how many electrons are in the various energy levels of their atoms. In fact, the electron arrangement of its atoms is what gives an element its chemical properties. One of the most important chemical properties of an element is its bonding (combining) ability. Some elements easily form bonds with other elements. Some elements hardly ever form bonds. An element's bonding ability is determined by the arrangement of the electrons in its atoms—more specifically, the arrangement of the electrons in the outermost energy level, or the level farthest from the nucleus.

Can the atom be "cut"? The existence of protons, neutrons, and electrons proves it can. In fact, two of these particles can be separated into even smaller particles. It is now believed that a new kind of particle makes up all the other known particles in the nucleus. This particle is called the **quark** (kwork). There are a number of different kinds of quarks. All nuclear particles are thought to be combinations of three quarks. One group of three quarks will produce a neutron. Another group of three quarks will produce a proton. According to current theory, quarks have properties called "flavor" and "color." There are six different flavors and three different colors.

4–2 Section Review

1. Classify the three main subatomic particles according to location, charge, and mass.
2. Why does the nucleus account for 99.9 percent of the mass of an atom?
3. Define atomic number; isotope; atomic mass; mass number.
4. Describe the arrangement of electrons in an atom. Why is electron arrangement so important?
5. Nitrogen-14 and nitrogen-15 are isotopes of the element nitrogen. Describe how atoms of these isotopes differ from each other.

Connection—*You and Your World*
6. How does a scientific model—such as the model of the structure of an atom—differ from a model airplane or boat that you might build?

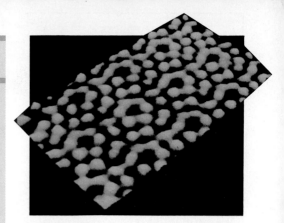

Figure 4–19 *This is the first image taken of atoms and their bonds. The bright round objects are single atoms. The fuzzy areas between atoms represent bonds.*

4–3 Forces Within the Atom

What keeps an atom together? Why don't the electrons fly out of their orbits around the nucleus? Why don't the protons move away from each other? Why don't all the atoms in the universe explode?

The answers to these questions can be found in the forces within the atom. **The four forces that account for the behavior of subatomic particles are the electromagnetic force, the strong force, the weak force, and gravity.**

The **electromagnetic force** can either attract or repel the particles on which it acts. If the particles have the same charge, such as two protons, the electromagnetic force is a force of repulsion. If the particles have opposite charges—such as an electron and a proton—the electromagnetic force is a force of attraction.

Electrons are kept in orbit around the nucleus by the electromagnetic force. The negatively charged electrons are attracted to the positively charged nucleus.

Guide for Reading

Focus on this question as you read.

▶ *What four forces are associated with atomic structure?*

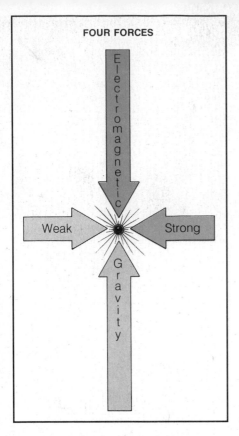

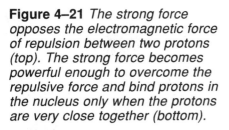

Figure 4–20 *The four known forces that govern all the interactions of matter and energy are the strong force, the electromagnetic force, the weak force, and gravity. Which of the four forces is the weakest?*

The electromagnetic force acts in the nucleus as a force of repulsion between positively charged protons. What keeps the protons from repelling each other and causing the atom to explode?

The **strong force** opposes the electromagnetic force of repulsion between protons. The strong force "glues" protons together to form the nucleus. Without the strong force, there would be no atoms. The strong force works only when protons are very close together, however. Although the strong force is the greatest of the four forces, it has a limited range. See Figure 4–21.

The **weak force** is the key to the power of the sun. The weak force is responsible for a process known as radioactive decay. During radioactive decay, a neutron in the nucleus changes into a proton and an electron.

The final force, **gravity,** is by far the weakest force known in nature. Yet it is probably the force most familiar to you. Gravity is the force of attraction exerted between all objects in nature. Gravity causes apples to fall from a tree and planets to remain in orbit around the sun. The effects of gravity are most easily observed in the behavior of large objects. Inside the nucleus of an atom, the effect of gravity is small compared to the effects of the other three forces. The role of gravity in the atom is not clearly understood.

As you can see, the four forces—electromagnetic, strong, weak, and gravity—are quite different. Yet physicists have tried to develop a single principle

Figure 4–21 *The strong force opposes the electromagnetic force of repulsion between two protons (top). The strong force becomes powerful enough to overcome the repulsive force and bind protons in the nucleus only when the protons are very close together (bottom).*

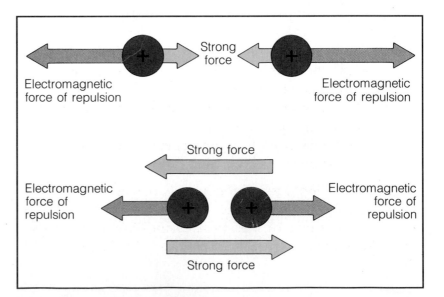

that would account for the differences between these forces. Such a principle would explain all four forces in terms of one fundamental force and all varieties of particles in terms of one basic particle. It is an awesome endeavor, indeed. But it is one that continues to challenge the knowledge and imagination of many scientists.

Figure 4–22 *In trying to unlock nature's secrets, physicists in a laboratory often study the tiny structure of the atom. Other scientists often find themselves in more precarious positions! These biologists are studying life in the treetops of a large tropical rain forest. As a result of their work, they may discover a previously unknown animal or plant, or even a new chemical that can be used to treat a once incurable disease. Often these new chemicals are analyzed atom by atom in a laboratory. Thus the knowledge uncovered in one scientific field often has applications in others.*

4–3 Section Review

1. What four forces govern the behavior of sub-atomic particles?
2. Which two forces are responsible for holding the atom together?
3. How does the electromagnetic force differ from the other three forces?

Critical Thinking—*Relating Concepts*
4. Gravity is the weakest of the four forces. However, it is one of the most easily observed forces in your daily life. Explain why this is so.

Laboratory Investigation

Shoe Box Atoms

Problem

How can indirect evidence be used to build a model?

Materials *(per group)*

| shoe box, numbered and taped shut, containing unidentified object(s) |
| balance |
| magnet |

Procedure

1. Your teacher will give you a shoe box with an object or objects inside. Do not open or damage the box.
2. Use a magnet to determine if the objects in the box have any magnetic properties.
3. Determine the mass of an empty shoe box. Then determine the mass of your shoe box. The difference between the two masses is the mass of the object(s) inside your shoe box.
4. You may be able to determine something about the object's shape by tilting the box. Does the object slide? (flat) Does it roll? (rounded) Does it collide inside? (more than one object)
5. Shake the box up and down to determine if the object bounces. How hard does it bounce? Does it flip?
6. For each test you perform, record your observations in a data table similar to the one shown here.

Observations

1. How many objects are in your shoe box?
2. Is the object soft? Magnetic? Fragile?
3. Is the object flat, or rounded?

Test Performed	Results	
	Trial 1	Trial 2
Magnet brought near		
Mass of object(s) determined		
Box tilted		
Box shaken		

Analysis and Conclusions

1. Make a sketch of what you think is in the shoe box. Draw the object(s) to show relative size.
2. What other indirect evidence did you gather to help you make the drawing?
3. How does your sketch compare with the actual objects as reported by your teacher? Make a sketch of the actual contents of the box.
4. Describe how you can develop a model of an object without directly observing the object.
5. **On Your Own** Prepare a shoe box model with two items that you select. Have a classmate see if he or she can determine what is in your shoe box.

Summarizing Key Concepts

4–1 An Atomic Model of Matter

▲ More than 2400 years ago, the Greek philosopher Democritus theorized the existence of the atom, the smallest particle of matter.

▲ John Dalton's atomic theory was based on experimental evidence about the behavior of matter. His theory stated that all matter is made of indivisible particles, or atoms.

▲ The discovery of the electron by J. J. Thomson proved that the atom is divisible.

▲ Thomson's model pictured the atom as being made of a positively charged, puddinglike material throughout which negatively charged electrons were scattered.

▲ Rutherford's experiments led him to propose an atomic model that states that an atom has a small, dense, positively charged nucleus surrounded by negatively charged electrons.

▲ The Bohr model of the atom pictured electrons as moving in definite orbits, or energy levels, around the nucleus.

▲ According to the theory of wave mechanics, electrons do not move about an atom in definite orbits. The exact location of an electron in an atom is impossible to determine.

4–2 Structure of the Atom

▲ Protons and neutrons are found in the nucleus.

▲ Protons have a positive charge and a mass of 1 amu.

▲ Neutrons are electrically neutral and have a mass of 1 amu.

▲ The number of protons in the nucleus of an atom is the atomic number.

▲ Atoms of the same element that have the same number of protons but different numbers of neutrons are called isotopes.

▲ The mass number of an atom is the sum of the protons and neutrons in its nucleus.

▲ The atomic mass of an element is the average mass of all the naturally occurring isotopes of that element.

▲ Electrons have a negative charge.

▲ Within the electron cloud, electrons are arranged in energy levels.

4–3 Forces Within the Atom

▲ Four forces—electromagnetic, strong, weak, and gravity—govern the behavior of subatomic particles.

Reviewing Key Terms

Define each term in a complete sentence.

4–1 An Atomic Model of Matter
nucleus

4–2 Structure of the Atom
subatomic particle
proton
atomic mass unit

neutron
atomic number
isotope
mass number
atomic mass
electron
electron cloud
energy level
quark

4–3 Forces Within the Atom
electromagnetic force
strong force
weak force
gravity

Chapter Review

Content Review

Multiple Choice

Choose the letter of the answer that best completes each statement.

1. The name Democritus gave to the smallest possible particle of matter is the
   a. molecule.   c. electron.
   b. atom.   d. proton.
2. The scientist J. J. Thomson discovered the
   a. proton.   c. neutron.
   b. electron. d. nucleus.
3. The small, heavy center of the atom is the
   a. neutron.   c. electron.
   b. proton.   d. nucleus.
4. Particles smaller than the atom are called
   a. molecules. c. ions
   b. elements. d. subatomic particles.
5. The nucleus of an atom contains
   a. protons and neutrons.
   b. protons and electrons.
   c. neutrons and electrons.
   d. protons, neutrons, and electrons.
6. The number of protons in an atom with an atomic number of 18 is
   a. 10.   b. 36.   c. 18.   d. 8.

7. An isotope of oxygen, atomic number 8, could have
   a. 8 protons and 10 neutrons.
   b. 10 protons and 10 neutrons.
   c. 10 protons and 8 electrons.
   d. 6 protons and 8 neutrons.
8. All nuclear particles are thought to be made of a combination of three
   a. electrons.   c. molecules.
   b. isotopes.   d. quarks.
9. Which of the following forces within the atom is responsible for keeping electrons in orbit around the nucleus?
   a. electromagnetic   c. weak
   b. strong   d. gravity
10. The arrangement and location of what subatomic particles determine the chemical properties of an atom?
   a. protons   c. quarks
   b. neutrons   d. electrons

True or False

If the statement is true, write "true." If it is false, change the underlined word or words to make the statement true.

1. Most of the mass of the atom is located in the electron cloud.
2. Electrons that have the least amount of energy are located farthest from the nucleus.
3. The idea that matter is made up of indivisible particles called atoms was proposed by Democritus.
4. In Thomson's experiment, the gas in the tube gave off rays that were made of negatively charged particles called neutrons.
5. The element chlorine has an atomic number of 17. It has 17 protons in its nucleus.

Concept Mapping

Complete the following concept map for Section 4–1. Refer to pages N6–N7 to construct a concept map for the entire chapter.

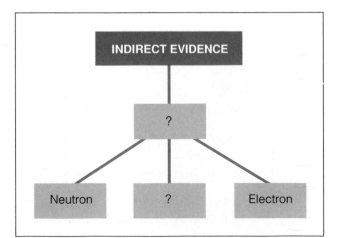

Concept Mastery

Discuss each of the following in a brief paragraph.

1. Describe the structure of the atom in terms of the three main subatomic particles. Include information about the location, charge, and atomic mass of each particle.
2. Describe the four forces and explain their role in the structure of the atom.
3. The model of the structure of the atom has changed over time. How does this illustrate the strength of the scientific method? Use specific examples in your explanation when appropriate.
4. Describe the electron configuration of each element based on atomic number: sulfur, 16; fluorine, 9; argon, 18; lithium, 3.
5. Explain how the following terms are related: atomic number, isotope, mass number, atomic mass. Be sure to define each term as you explain the relationships.
6. A certain element contains 82 percent of an isotope of mass number J and 18 percent of an isotope of mass number K. Is the atomic mass of this element closer to J or to K? Explain your answer.
7. Why must scientists use the concept of probability in describing the structure of an atom?
8. What is the significance of the atomic number of an element?

Critical Thinking and Problem Solving

Use the skills you have developed in this chapter to answer each of the following.

1. **Making inferences** Why are models useful in the study of atomic theory?
2. **Analyzing diagrams** The accompanying illustration shows the nucleus of a helium atom. Why is the nucleus positively charged? What is the mass of the nucleus? What is the atomic number of helium?

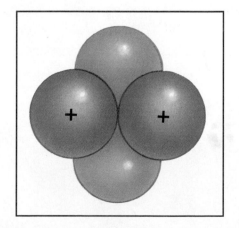

3. **Applying definitions** If the letter Z represents the atomic number of an atom and the letter A represents the mass number, explain how you could use these symbols to calculate: the number of protons, the number of electrons, and the number of neutrons.
4. **Making inferences** The element Einsteinium (named after the famous physicist Albert Einstein) has an atomic mass of 252. Einsteinium is a synthetic element, or an element that has been artificially made. How does this fact explain why the atomic mass of Einsteinium is a whole number?
5. **Using the writing process** Use the information presented in this chapter to write an update letter to Democritus. In your letter, explain how his early ideas about the atom have been modified as new pieces of indirect evidence have been uncovered. Be sure to include the names of the scientists who contributed to our current ideas about the atom and a description of their experiments.

Classification of Elements: The Periodic Table

Guide for Reading

After you read the following sections, you will be able to

5–1 Arranging the Elements
- Discuss the role Mendeleev played in the development of the periodic table.

5–2 Design of the Periodic Table
- Recognize how the modern periodic table is designed.
- Describe some differences between metals and nonmetals.

5–3 Chemical Families
- Describe some properties of eight families of elements.

5–4 Periodic Properties of the Elements
- Identify how the periodic law explains the physical and chemical properties of the elements.

Mary Cassatt, an American artist, lived in Paris toward the end of the nineteenth century. At that time artists were developing a "new" style of painting called Impressionism. Impressionist painters felt they no longer needed to paint with great realism. They used bright colors to give the "impression" of light and shadow in their paintings. Often, Impressionist painters moved out of their studios and into the daylight. Mary Cassatt began to paint in this new style. Today her paintings rightly share a place of honor in the history of human expression in art.

The origins of this painter's talents were firmly rooted in the human spirit. However, her paints' wondrous colors—colors that artists call pigments— have a much more ordinary origin. They are all part of the storehouse of elements in the Earth. A list of pigments reads like a chemistry textbook—zinc white, cadmium yellow, cobalt blue, iron oxide red, chromium green. All of these pigments are made from a group of elements known as the transition metals. In this chapter you will read about the transition metals and about many other groups of elements. Each group has its own properties and its own interesting—and often surprising—uses.

Journal *Activity*

You and Your World Artists often "speak" to people through drawings, paintings, and sculpture. Think back to the last time you were moved by a work of art. Maybe it was your own. Express your thoughts and feelings in your journal. Did you have any questions about what the artist was "saying"?

◀ *Mary Cassatt painted this picture entitled* Susan on a Balcony Holding a Dog.

5-1 Arranging the Elements

In some ways, your daily life contains many of the qualities of a good detective story. Unanswered questions and unexplained problems challenge your thoughts and actions all the time. And so you often act like a good detective. You discover places to go with friends and family. You learn about people who come from different cultures. You explore and gather evidence about the world around you. Like a good detective, you often try to make sense out of a series of clues and find meanings in seemingly unrelated observations.

Chemists, too, are good detectives. They gather evidence, analyze clues, follow their hunches, and make predictions. One of the most successful detective stories in the history of chemistry is the development of the periodic table of the elements. The periodic table of the elements represents a system of classifying, or logically grouping, all of the known elements. The arrangement of the elements in the periodic table was a milestone in the history of chemistry because it brought order to what had seemed to be a collection of thousands of unrelated facts. And it did something even more important: It helped chemists predict the existence of elements that had yet to be discovered!

Figure 5–1 *Each element has its own characteristic chemical and physical properties. Potassium is a soft, silvery metal that reacts explosively with water (right). Aluminum, also a silvery metal, does not easily combine with oxygen in water or in the air (left). Thus aluminum can be used for a variety of purposes, including drain pipes.*

A Hidden Pattern

The detective in this fascinating story was the Russian chemist Dmitri Mendeleev (duh-MEE-tree mehn-duh-LAY-uhf). The evidence he uncovered consisted of a huge collection of facts about the 63 elements that had been discovered by the mid-1800s. His clues were the physical and chemical properties of these elements. Based on these properties, it seemed clear to Mendeleev that some elements were similar to others. For example, sodium and potassium were both soft silver-white metals that reacted violently with water.

Mendeleev had a hunch that there had to be some order or relationship among all the elements. He was convinced that he could find a way to arrange the elements so that those with similar properties were grouped together. But what could the pattern of this arrangement be?

In his search for the pattern, Mendeleev first decided to organize his data. He did this by making a card for each of the known elements. On the card, he wrote the properties of each element. Some of the properties he included were atomic mass, density, color, and melting point. He also included the element's valence, or bonding power. Atoms form bonds with other atoms during chemical reactions (processes in which atoms join together to form molecules). When atoms form bonds, they either lose electrons, gain electrons, or share electrons. The valence, or valence number, of an element indicates the number of electrons that will be lost, gained, or shared in the bonding process.

Always looking for a pattern, Mendeleev arranged the cards in order of increasing atomic mass. If he started with lithium, the next element would be beryllium. Then would come boron, carbon, nitrogen, oxygen, and fluorine. With the cards arranged in this order, Mendeleev noticed the startling pattern of the valences: 1 2 3 4 3 2 1. Seven elements in a row, and the pattern of valences repeated itself.

As he arranged all 63 cards in order of atomic mass, Mendeleev saw the same pattern of rises and falls of valence again and again. He also saw something even more remarkable. When the elements were arranged in this way, they fell into columns,

Figure 5–2 *Mendeleev's greatest scientific contribution was the development of the periodic table. But his interests were not limited to chemistry. In 1887, he attempted to study a solar eclipse from a hot-air balloon.*

Figure 5–3 *The days of the month are periodic because they repeat themselves according to a definite pattern. What is that pattern? Some animals also behave in a periodic manner. Geese and other birds migrate every fall and spring. What other examples of periodic behavior have you observed?*

one under the other. All the elements in a column had the same valence! All the elements in a column showed similar physical and chemical properties!

It was obvious to Mendeleev that the properties of the elements recurred at regular intervals. In Mendeleev's words, he found that "the properties of the elements were periodic functions of their atomic masses." When used this way, the word periodic means repeating according to some pattern. The days of the week are periodic because every seven days the pattern recurs. The months of the year are periodic because they also occur in a regular, repeating pattern. The notes of the musical scale are periodic, repeating a pattern with every eighth tone. In fact, you may already understand the word periodic from your familiarity with the word periodical. Sometimes magazines and newspapers are called periodicals. Their appearance on a newsstand, in a library, or at your home occurs according to a recognized repeating pattern. (You probably know when your favorite periodical is due to appear and eagerly anticipate its arrival.) Animals and plants also exhibit periodic behaviors. Birds fly south when winter's cold limits their food supply, and they return north the following spring when warmth signals an abundance of food. Can you think of other examples of periodic occurrences?

A Bold Prediction

Mendeleev designed a periodic table in which the elements were arranged in order of increasing atomic mass. Confident of the accuracy of his discovery, he left spaces in the table in order to make the known elements fit into the proper columns. Then he boldly announced that the empty spaces would be filled with elements that were not yet discovered! Indeed, he even went so far as to predict the physical and chemical properties of the unknown elements. He based his predictions on the properties of the elements above and below and to the left and right of the spaces in the table. Was he correct?

Yes, in fact, he was. Three of the unknown elements were discovered and placed in their correct positions in the periodic table during his lifetime. And the properties of the newly discovered elements

MENDELEEV'S PREDICTIONS AND ACTUAL PROPERTIES OF ELEMENT 32

"Ekasilicon"		Germanium	
Date predicted	1871	Date discovered	1886
Atomic mass	72	Atomic mass	72.6
Density	5.5 g/cm^3	Density	5.47 g/cm^3
Bonding power	4	Bonding power	4
Color	Dark gray	Color	Grayish white

Figure 5–4 *The discovery of the element germanium in 1886 made Mendeleev the most famous chemist of the time. Notice how his predictions about the properties of element 32, or "ekasilicon," were incredibly close to the actual properties. How could Mendeleev predict the properties of an "unknown" element with such accuracy?*

were in close agreement with Mendeleev's predictions. You can see for yourself how well Mendeleev's predictions were fulfilled by looking at Figure 5–4.

The Modern Periodic Table

Despite the importance of Mendeleev's work, his periodic table was not perfect. When the elements are arranged in order of increasing atomic mass, several elements appear to be misplaced in terms of their properties. Mendeleev assumed that this was because the atomic masses of these elements had been incorrectly measured. Yet new measurements reconfirmed the original masses. What could be the problem?

It was not until 50 years after Mendeleev had developed his periodic table that the answer to the problem became clear. It was then that the British scientist Henry Moseley determined for the first time the atomic numbers of the elements. As you will recall, the atomic number of an element is the number of protons in the nucleus of each atom of that element.

The discovery of atomic numbers led to an important change in Mendeleev's periodic table. It turns out that when the elements are arranged in order of increasing atomic number (rather than increasing atomic mass), elements with similar physical and chemical properties fall into place without exception. Thus, Mendeleev's periodic table was replaced by the modern periodic table. The **periodic law** forms the basis of the modern periodic table.

Figure 5–5 *Great contributions are often made by the young. Henry Moseley was only 27 when he died in a famous battle during World War I, but his work in developing the modern periodic table lives long after him. What was Moseley's contribution?*

Figure 5–6 *Mendeleev recognized that the properties of elements are repeated in a periodic way. Thus certain elements have similar properties. Silver (top), gold (center), and copper (bottom) are all shiny, hard elements that are good conductors of electricity. What are some other uses of these elements?*

The periodic law states that the physical and chemical properties of the elements are periodic functions of their atomic numbers.

Just in case you have the impression that all scientists are strange old people working mysteriously in musty laboratories, you might be interested to learn that Henry Moseley completed his historic work before his twenty-eighth birthday. Sadly, he died during World War I at the battle of Gallipoli. It must be left to our imaginations to wonder what other contributions to human knowledge this brilliant chemist might have made had he survived.

5–1 Section Review

1. Describe Mendeleev's periodic table. What are some properties he used to order the elements?
2. What does the word periodic mean?
3. How did Mendeleev predict the existence of undiscovered elements?
4. According to the modern periodic law, what determines the order of the elements?

Connection—*You and Your World*
5. An important series of reference books found in most libraries is entitled *Readers' Guide to Periodical Literature.* From the title, can you infer what kinds of information can be found in these books? Why do you think this series of books is important?

5–2 Design of the Periodic Table

The periodic table of the elements is one of the most important tools of a scientist, especially a chemist. Why? Because the periodic table is a classification system—a way of organizing vast amounts of information in a logical, usable, and meaningful way. In some ways the periodic table is like the system used to organize books in a library. Imagine how confusing it would be if all the books in a library were placed on shelves in no particular order. You would have a hard time trying to locate the book on magic tricks, stamp collecting, or rocketry you were looking for.

Fortunately, the books in a library are arranged by subject in a system that uses numbers and letters. You can look up a book in the card catalog, find its classification number, and locate it on the shelves. As you can see, such organization makes a book easy to locate. There is another advantage to a library's classification system. The classification number is a key to a book's subject matter. It identifies the broad topic of a book. All books with the same or similar subject matter will have essentially the same classification number. So without ever having seen a certain book, you can predict its topic from its classification number or its placement on a shelf.

The periodic table organizes the elements in a particular way. A great deal of information about an element can be gathered from its position in the periodic table. For example, you can predict with reasonably good accuracy the physical and chemical properties of the element. You can also predict what other elements a particular element will react with chemically. This means that it is not necessary to memorize a whole list of facts about many different elements. Understanding the organization and plan of the periodic table will help you obtain basic information about each of the 109 known elements. A periodic table is found on pages 114 and 115. Refer to the table often as you read about it.

Columns in the Periodic Table

If you look at the periodic table in Figure 5–8, you will notice that it consists of vertical columns of elements. Each column is numbered. There are eighteen main columns of elements. Columns of elements are called **groups** or **families**. Elements within the same group or family have similar but not identical properties. For example, lithium (Li), sodium (Na), potassium (K), and the other members of Family 1 are all soft, white, shiny metals. They are all highly reactive elements, which means they readily combine with other elements to form compounds.

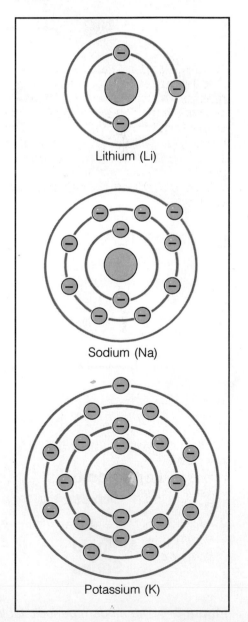

Figure 5–7 *Elements in the same family of the periodic table have similar properties. Here you see the electron arrangement of the elements lithium, sodium, and potassium. How is the electron arrangement in each element similar?*

Lithium (Li)

Sodium (Na)

Potassium (K)

PERIODIC TABLE

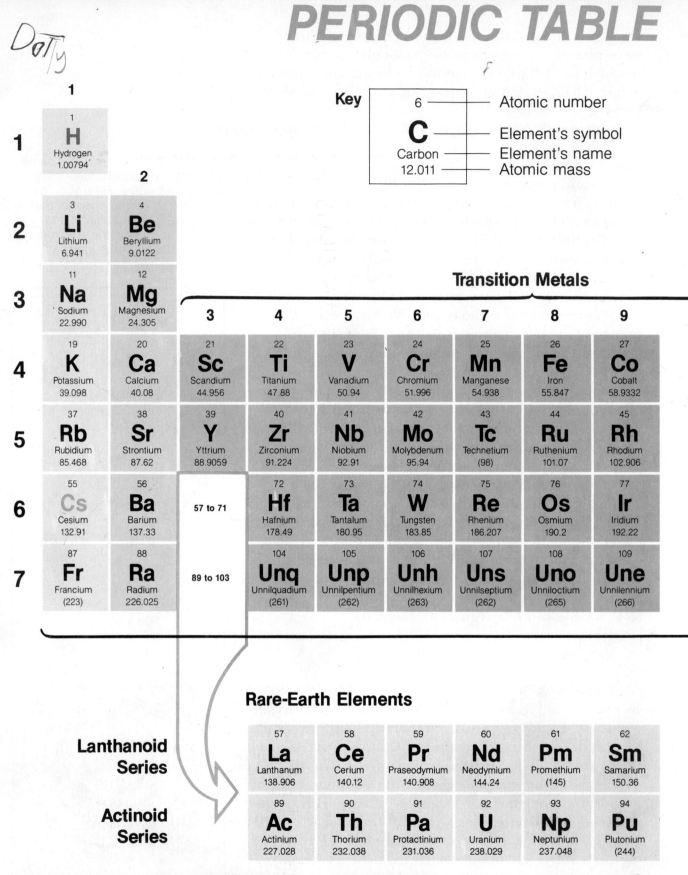

Key

6	Atomic number
C	Element's symbol
Carbon	Element's name
12.011	Atomic mass

Transition Metals

Figure 5–8 *The modern periodic table of the elements is shown here.*

OF THE ELEMENTS

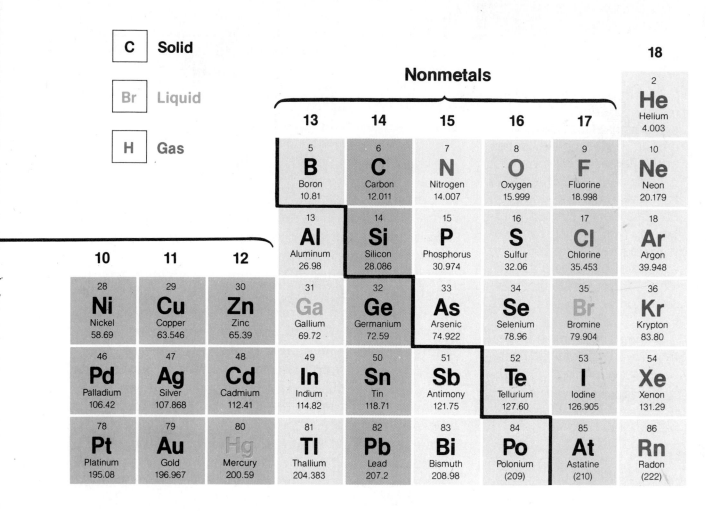

C	Solid
Br	Liquid
H	Gas

Nonmetals

18

					18
13	14	15	16	17	2 **He** Helium 4.003

| 10 | 11 | 12 | 13 **Al** Aluminum 26.98 | 14 **Si** Silicon 28.086 | 15 **P** Phosphorus 30.974 | 16 **S** Sulfur 32.06 | 17 **Cl** Chlorine 35.453 | 18 **Ar** Argon 39.948 |

| 5 **B** Boron 10.81 | 6 **C** Carbon 12.011 | 7 **N** Nitrogen 14.007 | 8 **O** Oxygen 15.999 | 9 **F** Fluorine 18.998 | 10 **Ne** Neon 20.179 |

28 **Ni** Nickel 58.69	29 **Cu** Copper 63.546	30 **Zn** Zinc 65.39	31 **Ga** Gallium 69.72	32 **Ge** Germanium 72.59	33 **As** Arsenic 74.922	34 **Se** Selenium 78.96	35 **Br** Bromine 79.904	36 **Kr** Krypton 83.80
46 **Pd** Palladium 106.42	47 **Ag** Silver 107.868	48 **Cd** Cadmium 112.41	49 **In** Indium 114.82	50 **Sn** Tin 118.71	51 **Sb** Antimony 121.75	52 **Te** Tellurium 127.60	53 **I** Iodine 126.905	54 **Xe** Xenon 131.29
78 **Pt** Platinum 195.08	79 **Au** Gold 196.967	80 **Hg** Mercury 200.59	81 **Tl** Thallium 204.383	82 **Pb** Lead 207.2	83 **Bi** Bismuth 208.98	84 **Po** Polonium (209)	85 **At** Astatine (210)	86 **Rn** Radon (222)

The symbols shown here for elements 104-109 are being used temporarily until names for these elements can be agreed upon.

Metals

Mass numbers in parentheses are those of the most stable or common isotope.

63 **Eu** Europium 151.96	64 **Gd** Gadolinium 157.25	65 **Tb** Terbium 158.925	66 **Dy** Dysprosium 162.50	67 **Ho** Holmium 164.93	68 **Er** Erbium 167.26	69 **Tm** Thulium 168.934	70 **Yb** Ytterbium 173.04	71 **Lu** Lutetium 174.967
95 **Am** Americium (243)	96 **Cm** Curium (247)	97 **Bk** Berkelium (247)	98 **Cf** Californium (251)	99 **Es** Einsteinium (254)	100 **Fm** Fermium (257)	101 **Md** Mendelevium (258)	102 **No** Nobelium (259)	103 **Lr** Lawrencium (260)

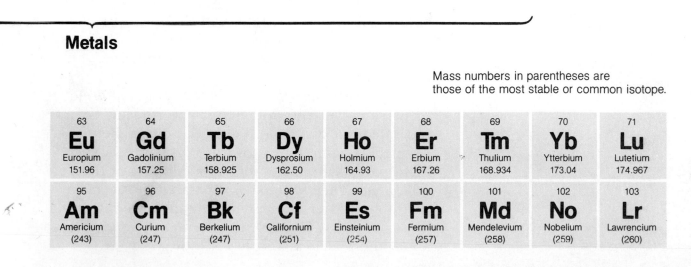

ACTIVITY

DISCOVERING

Classifying Objects—My Way

Mendeleev's table and the modern periodic table are systems of classifying the elements based on similarities and differences in properties.

Choose a set of objects that are familiar to you. You can use coins, stamps, marbles, leaves, playing cards, or jelly beans. Devise your own system of classifying the objects. Put your system of classification in a table for others to use.

■ What is the most important advantage of your classification system?

Fluorine (F), chlorine (Cl), bromine (Br), and iodine (I) are four of the elements in Family 17. Although fluorine and chlorine are both gases, bromine is a liquid, and iodine is a solid, they all still have many similar properties. In fact, both bromine and iodine become gases very easily. All four elements react to form the same kinds of compounds. You will learn more about each family and its properties later in this chapter.

Rows in the Periodic Table

As you look at the periodic table once again, observe that each horizontal row of elements is called a **period.** Unlike the elements in a family, the elements in a period are not alike in properties. In fact, the properties of the elements change greatly across any given row.

But there is a pattern to the properties of the elements as one moves across a period from left to right. The first element in a period is always an extremely active solid. The last element in a period is always a particularly inactive gas. You can see this pattern by looking at the elements in Period 4 of the periodic table. The first element, potassium (K), is an active solid. The last element, krypton (Kr), is an inactive gas (and bears no relationship to the fictional element Kryptonite, which is the only thing feared by Superman!). The symbols for the elements potassium and krypton should remind you of a rule for writing chemical symbols that you learned about in Chapter 3. The chemical symbol for an element consists of one or two letters. If it consists of one letter, the letter is always capitalized. If it consists of two letters, the first letter is always capitalized, but the second never is.

As you can see, there are seven periods of elements. You will also notice that one row has been separated out of Period 6 and one out of Period 7. Even though these two rows are displayed below the main part of the table, they are still part of the periodic table. They have been separated out to make the table shorter and easier to read. Elements in these two rows are rare-earth elements. You will read about these elements in just a little while.

Element Key

Look closely at the periodic table. Each element is found in a separate square. **Important information about an element is given in each square of the periodic table: its atomic number, chemical symbol, name, and atomic mass.**

The number at the top of each square is the atomic number of the element. Remember that the atomic number of an element is the number of protons in an atom of that element. The atomic number is unique to that element. In other words, no two elements have the same atomic number. Look closely at the element squares of the periodic table to see for yourself that this is true. And as you're looking, notice that the elements are arranged in order of increasing atomic number.

Just below the atomic number, near the center of the square, is the chemical symbol for the element. Below the chemical symbol, the name is spelled out. The number near the bottom of the square is the atomic mass of the element.

Figure 5–9 *The properties of elements in the same period are not alike. Reading from left to right are the elements potassium, iron, copper, gallium, and bromine. In what ways do the properties change across the period?*

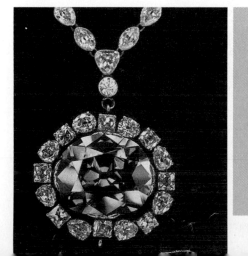

6

C

Carbon

12.011

Figure 5–10 *Four important facts about an element are supplied in each square of the periodic table: the atomic number, symbol, name, and atomic mass of that element. The element carbon is found in many things. It makes up the lead in your pencil, most of the foods you eat, and also the Hope diamond.*

A Hot Time

1. Roll a piece of aluminum foil into a small cylinder about the shape of a pencil.

2. Stand a stainless steel teaspoon, a strip of wood, a plastic spoon, and the aluminum cylinder you made in a plastic cup.

3. Add hot water to the cup. Be careful to leave the tops of the objects above the water level.

4. Wait one minute. Carefully touch the exposed ends of each object in the cup. What are your observations?

■ What conclusions does this activity help you discover?

Figure 5–11 *Unlike some metals, pure gold is actually quite soft and easily worked. This gold pin, hammered from a flat sheet of almost pure gold, can be bent out of shape with little effort. Gold used for jewelry is often combined with copper and other metals to give it strength.*

Now practice using what you have just learned. Locate the element boron in the periodic table. What is its atomic number? Its symbol? What element has the symbol Cd? What element has an atomic number of 38? What is the atomic mass of magnesium? Of bromine?

Metals

When you hear the word **metal,** what do you think of? You probably think of silver, iron, or copper. These are all familiar metals. Kitchen pots, trumpets, knives and forks, and pocket change are all familiar items made of metals. However, there are other elements that are classified as metals that you may not have thought of as metals—such as calcium, sodium, and potassium.

Of the 109 known elements, most are metals. If you look at the periodic table, you will see a dark zigzag line running like steps down the right side of the table. The 88 elements to the left of this line are metals or metallike elements.

PHYSICAL PROPERTIES OF METALS The physical properties of metals make them easy to identify. One such property is **luster,** or shininess. Hold a brand-new penny in your hand, and its gleam will convince you of this important property of metals. Most metals also allow heat and electricity to move through them easily. Therefore, metals are good conductors of heat and electricity. In general, metals have a high density. This means that they are heavy for their size. Trying to lift a metal dumbbell can easily convince you of this! Finally, metals usually have fairly high melting points. Now that you know about some of the properties of metals, explain why metal pots are used for cooking.

There are two other physical properties that are common to many metals. Most metals are **ductile,** which means they can be drawn out into thin wires. And most metals are **malleable,** which means they can be hammered into thin sheets. The ease with which metals can be drawn into wire and hammered into sheets contributes, in large part, to their use in making jewelry. The pin in Figure 5–11 is made of a thin gold sheet that was hammered by an ancient Peruvian native.

Figure 5–12 *Some of the physical properties of metals and their alloys are evident in these photographs. What properties can you identify in the cooling slabs of steel (left), steel girders (center), stainless steel artificial human hip joint (inset), and copper tubes (right)?*

CHEMICAL PROPERTIES OF METALS The chemical properties of metals are not as easily observed as are the physical properties. The chemical properties of any element depend upon the electron arrangement in the atoms of the element—more specifically, on the number of electrons in the outermost energy level. (Remember that the electrons in the outer energy level, or the valence electrons, are involved in forming bonds with other atoms.) An atom of a metal can have 1, 2, 3, or 4 electrons in its outermost energy level. The electrons in the outermost energy level of a metal are rather weakly held. So metals are elements that tend to lose their outermost electrons when they combine chemically.

Because they tend to lose electrons, most metals will react chemically with water or with elements in the atmosphere. Such a chemical reaction often results in **corrosion.** Corrosion is the gradual wearing away of a metal due to a chemical reaction in which the metal element is changed into a metallic compound. The rusting of iron is an example of corrosion. When iron rusts, it combines with oxygen in the air to form the compound iron oxide. The

Figure 5–13 *The rusting of iron and steel is actually a form of corrosion. This abandoned car shows the dramatic effects of corrosion.*

tarnishing of silver is another example of corrosion. Tarnishing results when silver reacts with compounds of sulfur in the air or in certain foods. Have you ever observed examples of rusting and/or tarnishing? How would you describe what you observed?

Nonmetals

Elements that are **nonmetals** are located to the right of the zigzag line in the periodic table. Fewer elements are classified as nonmetals than as metals in the periodic table. In general, the physical and chemical properties of nonmetals tend to be opposite those of metals.

PHYSICAL PROPERTIES OF NONMETALS Nonmetals usually have no luster and are dull in appearance. Nonmetals do not conduct heat and electricity well. Nonmetals are brittle and thus break easily. They cannot be drawn out into wire or hammered into thin sheets. In other words, nonmetals are neither ductile nor malleable. Nonmetals usually have lower densities and lower melting points than metals.

Nonmetals are not as easy to recognize as a group as are metals. Nonmetals can be noticeably different from one another. For example, bromine is a brown liquid, oxygen is a colorless gas, and sulfur is a yellow solid. Yet all are nonmetals.

CHEMICAL PROPERTIES OF NONMETALS Remember, the chemical properties of elements are determined by the number of electrons in the outermost energy level. Atoms of most nonmetals have 5, 6, 7, or 8 electrons in their outermost energy level. Atoms with 5, 6, or 7 valence electrons gain 3, 2, or 1 electron, respectively, when they combine chemically. Thus nonmetals are elements that tend to gain electrons. Perhaps you are wondering about the nonmetals whose atoms have 8 valence electrons. Atoms with 8 valence electrons have a complete outermost energy level. So elements whose atoms have 8 valence electrons tend to be nonreactive or rarely react with atoms of other elements. Knowing what you do now about metals and nonmetals, do you think they can form compounds with each other if one gives up electrons and one takes electrons?

Figure 5–14 *Sulfur (top) is a nonmetal that can form beautiful crystals. Boron (bottom) is a metalloid, a word that means metallike. What are some properties of nonmetals and metalloids?*

Figure 5–15 *Both silicon (left) and antimony (right) are metalloids. All metalloids are solids that can be shiny or dull. The silicon in the photograph has been made into a computer chip.*

Metalloids

The dividing line between metals and nonmetals is not quite as definite as it appears. For along both sides of the dark zigzag line are elements that have properties of both metals and nonmetals. These elements are called **metalloids** (MEHT-uh-loidz). The word metalloid means metallike. All metalloids are solids that can be shiny or dull. They conduct heat and electricity better than nonmetals but not as well as metals. Metalloids are ductile and malleable. The metalloids include boron, silicon, germanium, arsenic, antimony, tellurium, polonium, and astatine. Do any of these elements sound familiar to you? If so, in what way?

5–2 Section Review

1. What important information is given in each square of the periodic table?
2. What are the horizontal rows in the periodic table called? What are the vertical rows called?
3. What are some physical and chemical properties of metals? Of nonmetals?
4. What is a metalloid?

Critical Thinking—*Applying Concepts*
5. Electricity is, in large part, responsible for our modern lifestyle. How are some of the physical properties of metals related to the fact that a steady supply of electricity is able to reach our homes from electric generating plants?

ACTIVITY
DOING

An Elemental Hunt

1. Collect some samples of elements that are easily obtained, such as copper, iron, aluminum, nickel, and carbon.

2. Attach each element to a large square of paper or cardboard. Make each square of paper or cardboard look like a square of the periodic table.

3. On the paper square, include the atomic mass, atomic number, name, and chemical symbol of the element displayed.

Working along with your classmates, see how complete a periodic table you are able to create.

Guide for Reading

Focus on these questions as you read.

▶ What is the basis for the placement of elements in the periodic table?

▶ What are the properties of chemical families?

5–3 Chemical Families

You are about to go on a "tour" of the periodic table. This trip will help you become more familiar with the basic properties of the families of elements. Remember, it is not necessary to memorize lots of facts about the elements. What we hope you will be able to do is recognize the value of the periodic table in organizing information about the elements. We also hope that you will appreciate how these elements—some of which may be unfamiliar to you—are a part of your world. After all, your world is made up of matter, and matter is ultimately made up of atoms of elements. So in some way, each and every one of the 109 elements in the periodic table is a part of your life. To help you on your tour, keep in mind the following principle: **Elements within the same family of the periodic table have similar properties because they have the same number of valence electrons.**

The Most Active Metals

The elements in Family 1, with the exception of hydrogen, are called the **alkali** (AL-kuh-ligh) **metals.** Atoms of the alkali metals have a single electron in their outermost energy level. In other words, they have 1 valence electron. Hydrogen also has 1 electron in its outer shell. In many ways, it behaves like the alkali metals.

The alkali metals are soft, silver-white, shiny metals. They are so soft, in fact, that they can be cut with a knife. The alkali metals are good conductors of heat and electricity. Because they have only 1 valence electron, these elements bond readily with other substances. In fact, they are so reactive that they are never found uncombined in nature. In other words, they are never found as free elements. In the laboratory, samples of these elements are stored in oil in order to keep them from combining with water or oxygen in the air. The reaction is violent when the alkali metals react with water. Hydrogen gas is produced, as well as extreme heat. Because of the heat produced, the hydrogen gas can begin to burn and may explode.

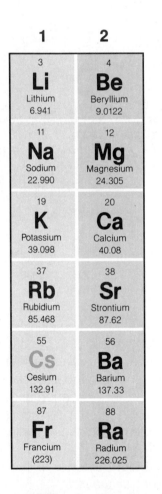

Figure 5–16 *Family 1 metals are called the alkali metals. Family 2 metals are called the alkaline earth metals. How many electrons are in the outermost energy level of each alkali metal? Of each alkaline earth metal?*

ALKALI METALS

Element	Properties		Uses of Compounds
Lithium (Li)	m.p. 179°C b.p. 1336°C		Medicine; metallurgy
	Soft; silvery; reacts violently with water		
Sodium (Na)	m.p. 97.8°C b.p. 883°C		Soap; table salt; lye
	Soft; silvery white; reacts violently with water		
Potassium (K)	m.p. 62.5°C b.p. 758°C		Fertilizer; medicine; photography
	Soft; silvery white; reacts violently with moisture		
Rubidium (Rb)	m.p. 39.0°C b.p. 700°C		Space vehicle engines; photocells
	Soft; lustrous; reacts violently with moisture		
Cesium (Cs)	m.p. 28.6°C b.p. 670°C		Photocells
	Silvery white; ductile; reacts with moisture		
Francium (Fr)	m.p. (27°C) b.p. (677°C)		Not widely used
	Extremely rare; radioactive isotopes		

Values in parentheses are physical properties of the most stable isotope.

Figure 5–17 *This table shows some of the properties of the alkali metals and the uses of their compounds. Which alkali metal has radioactive isotopes?*

Although the alkali metals themselves have few familiar uses, the compounds they form are some of the most important substances you use every day. Table salt and baking soda are two compounds you may be familiar with. Soap, which forms when alkali compounds react with fats, is another. Now look at the periodic table and identify the alkali metals.

Family 2 consists of the six elements known as the **alkaline** (AL-kuh-lihn) **earth metals.** Like the alkali metals, the alkaline earth metals are never found in nature as uncombined elements. Instead, they exist bonded with other elements in compounds. The alkaline earth metals have 2 valence electrons. Atoms of these elements lose their 2 electrons easily when they combine with other atoms. But since they must lose 2 electrons, they are not quite as reactive as the alkali metals.

Two of the alkaline earth metals—magnesium and calcium—are probably familiar to you. Magnesium is often combined with aluminum to make alloys that are strong yet light in weight. These alloys are used to make ladders and airplane parts. They are used where light yet strong metal parts are needed. Other compounds of magnesium are used in medicines, flares, and fireworks. Calcium is an abundant substance in the Earth's crust. Calcium compounds

ALKALINE EARTH METALS

Element	Properties	Uses of Compounds
Beryllium (Be)	m.p. 1285°C b.p. 2970°C Poisonous	Radio parts; steel
Magnesium (Mg)	m.p. 650°C b.p. 1117°C Burns with very bright flame; strong but not dense	Medicine; photographic flashbulbs; auto parts; space vehicle parts; flares
Calcium (Ca)	m.p. 851°C b.p. 1487°C Silvery; important part of bones and teeth; tarnishes in moist air	Plaster and plasterboard; mortar and cement; water softeners; metal bearings
Strontium (Sr)	m.p. 774°C b.p. 1366°C Least abundant alkaline earth metal; reactive in air	Fireworks; flares
Barium (Ba)	m.p. 850°C b.p. 1537°C Extremely reactive in air	Medicine; paints; glassmaking
Radium (Ra)	m.p. (700°C) b.p. (1525°C) Silvery white but turns black in air; radioactive	Treatment of cancer; medical research

Values in parentheses are physical properties of the most stable isotope.

Figure 5–18 *The properties and uses of alkaline earth metals are shown in this table. Which alkaline earth metal is important for strong teeth and bones?*

make up limestone and marble rock. Calcium is also an essential part of your teeth and bones.

The Transition Metals

Look at the periodic table between Family 2 and Family 13. What do you see? You should see several groups of elements that do not seem to fit into any other family. These elements are called the **transition metals.** Transition metals have properties similar to one another and to other metals, but they are different from the properties of any other family.

The names of the transition metals are probably well known to you. These are the metals with which you are probably most familiar: copper, tin, zinc, iron, nickel, gold, and silver, for example. You may

Figure 5–19 *An artist uses a palette of different colors to paint a picture. Many of the colors are made from the transition metals. Tungsten is a transition metal whose importance in your life is immediately apparent when you switch on an incandescent light. The tungsten wire in a light bulb glows as electricity passes through it.*

also know that the transition metals are good conductors of heat and electricity. The compounds of transition metals are usually brightly colored and are often used to color paint. (Remember Mary Cassatt and the other Impressionist painters you read about at the beginning of the chapter?) Gold and silver are used to make jewelry and eating utensils. These two metals are often used in dental fillings to replace decayed areas of a tooth. Silver is essential in the making of photographic film and paper. Mercury is an interesting transition metal because it is a liquid at temperatures above $-38.8°C$. How do you think this fact relates to the use of mercury in household thermometers?

Most transition elements have 1 or 2 valence electrons. When they combine with other atoms, they lose either 1 or both of their valence electrons. But transition elements can also lose an electron from the next-to-outermost energy level. In addition, transition elements can share electrons when they form bonds with other atoms. It is no wonder that transition elements form so many different compounds!

From Metals to Nonmetals

To the right of the transition elements are six families, five of which contain some metalloids. This means that certain members of these families show properties of metals as well as nonmetals. These four families are named after the first element in the family.

Family 13 is the **boron family.** Atoms of elements in this family have 3 valence electrons. Boron is a metalloid. The other elements, including aluminum, are metals.

Boron, which is hard and brittle, is never found uncombined in nature. It is usually found combined with oxygen. Compounds of boron are used to make heat-resistant glass, such as the test tubes used in your laboratory and the glass cookware used in your kitchen. Boron is also found in borax, a cleaning compound that may be familiar to you.

Aluminum is the most abundant metal and the third most abundant element in the Earth's crust. Aluminum is also found combined with oxygen in the ore bauxite. Aluminum is an extremely important

TRANSITION ELEMENTS

Element	Uses
Iron (Fe)	Manufacturing; building materials; dietary supplement
Cobalt (Co)	Magnets; heat-resistant tools
Nickel (Ni)	Coins; batteries; jewelry; plating
Copper (Cu)	Electric wiring; plumbing; motors
Silver (Ag)	Jewelry; dental fillings; mirror backing; electric conductor
Gold (Au)	Jewelry; base for money systems; coins; dentistry
Zinc (Zn)	Paints; medicines; coat metals
Cadmium (Cd)	Plating; batteries; nuclear reactors
Mercury (Hg)	Liquid in thermometers, barometers, electric switches; dentistry; paints

Figure 5–20 *The transition elements have many common uses. Which transition element is liquid at room temperature?*

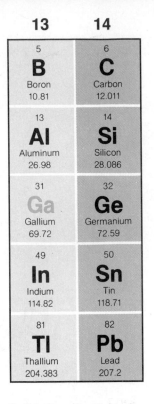

13	14
5 **B** Boron 10.81	6 **C** Carbon 12.011
13 **Al** Aluminum 26.98	14 **Si** Silicon 28.086
31 **Ga** Gallium 69.72	32 **Ge** Germanium 72.59
49 **In** Indium 114.82	50 **Sn** Tin 118.71
81 **Tl** Thallium 204.383	82 **Pb** Lead 207.2

Figure 5–21 *Family 13 is also known as the boron family. Elements in Family 14 are also known as the carbon family.*

Figure 5–22 *Carbon is found in oil, gas, and other petroleum products. In an oil refinery, crude oil is processed into many different products. Silicon is one of the most abundant elements in the Earth's crust.*

metal in industry. It is light, strong, and does not corrode. It is an excellent reflector of light and a good conductor of heat and electricity. Aluminum is used to make parts for cars, trains, and planes. It is also made into the pots and pans used in cooking. Because aluminum is so malleable, it can also be made into foil used to wrap food for storage.

The **carbon family** is Family 14. Atoms of the elements in this family have 4 valence electrons. Carbon is a nonmetal. But the next two elements, silicon and germanium, are metalloids. And tin and lead are metals.

Carbon atoms, with their 4 valence electrons, form an unbelievable number of different compounds—more than 5 million by some estimates! The element carbon is often called the "basis of life." Your body contains a wide variety of carbon compounds. Sugars and starches are two important examples. Fuels such as gasoline also contain carbon compounds. Carbon compounds are so numerous, in fact, that a whole branch of chemistry is devoted to their study. This branch is called organic chemistry.

Silicon is the second most abundant element in the Earth's crust. Silicon combined with oxygen to form sand is used to make glass and cement. Silicon is also used to make solar cells, which are able to convert the energy of sunlight into electricity. Solar cells are commonly found in roof panels and are also used on space satellites. One of the most important uses of silicon is probably quite familiar to you. Silicon chips are used for circuitry and memory in computers.

Figure 5–23 *Nitrogen is used to make explosives. This building is being demolished by a controlled explosion that causes the building to collapse.*

Germanium is a metalloid commonly used in transistors. Transistors are components of many electronic devices, including radios, televisions, and computer games. Tin is a metal that resists rusting and corrosion. The common "tin can" used as a container for your favorite soup is actually a steel can lined with a very thin layer of tin. The tin lining prevents the food in the can from coming into contact with the steel wall of the can.

Lead is another metal in the carbon family. In the past, lead was used to color paint. It was also an important additive in gasoline. However, because of the dangers associated with exposure to lead, it has been removed from paints and gasoline. And just so you do not become concerned about using an ordinary pencil, you should know that the "lead" in a lead pencil is not really lead but is a form of carbon known as graphite. A lead pencil is perfectly safe to use.

The **nitrogen family,** Family 15, is named after an element that makes up 78 percent of the air around you: nitrogen. The atoms of elements in this family have 5 valence electrons in their outermost energy level. These atoms tend to share electrons when they bond with other atoms.

Nitrogen is the most abundant element in the Earth's atmosphere. It is an exceptionally stable element and does not combine readily with other elements. Nitrogen is an important part of many fertilizers, which are substances used to enhance plant growth. Nitrogen is also used to produce explosives, medicines, and dyes. Ammonia, a common household cleaning agent, is a compound made of nitrogen and hydrogen.

Phosphorus is an active nonmetal that is not found free in nature. One of its main uses is in making the tips of matches. It is also used in flares. Arsenic is an important ingredient in many insecticides. Both antimony and bismuth are used in making alloys.

The elements making up Family 16 are called the **oxygen family.** Atoms of these elements have 6 valence electrons. Most elements in this family share electrons when forming compounds.

Activity Bank

What Is the Effect of Phosphates on Plant Growth?, p.159

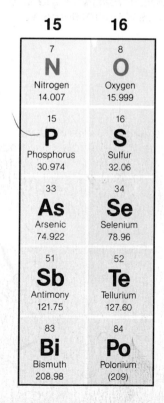

Figure 5–24 *Which element in Family 15 shows the most metallic properties? Which member of Family 16 is a gas?*

HALOGENS	
Element	**Uses**
Fluorine (F)	Etching glass; refrigerants; nonstick utensils; preventing tooth decay
Chlorine (Cl)	Bleaching agent; disinfectant; water purifier
Bromine (Br)	Medicine; dyes; photography
Iodine (I)	Medicine; disinfectant; dietary supplement in salt
Astatine (At)	Rare element

Figure 5–25 *Because of their chemical reactivity, the halogens have many uses.*

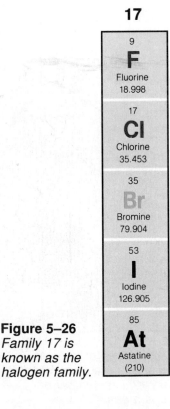

Figure 5–26 *Family 17 is known as the halogen family.*

Oxygen, the most abundant element in the Earth's crust and the second most abundant element in the atmosphere, is an extremely reactive element. It combines with almost every other element. You already know how important oxygen is to you—and to almost all other forms of life on Earth. Your body uses the oxygen you breathe to break down carbohydrates to produce energy. Plants also use oxygen to break down carbohydrates. In addition to combining with other elements, oxygen can also form molecules by bonding with itself. You might be familiar with the word ozone. Three atoms of oxygen bond to form a molecule of ozone, O_3. In the atmosphere, the ozone layer screens out harmful ultraviolet radiation from the sun, thus protecting life on Earth.

Sulfur, selenium, and tellurium are brittle solids at room temperature. They all combine with oxygen as well as with metals and with hydrogen. Sulfur is used to manufacture medicines, matches, gunpowder, and synthetic rubber. Selenium is used to color glass red and to make enamels. Tellurium is useful in making alloys. Polonium, another member of this family, is an extremely rare element.

The Halogens

The elements of Family 17 are fluorine, chlorine, bromine, iodine, and astatine. Together they are known as the **halogen family**. Halogens have 7 valence electrons, which explains why they are the most active nonmetals. Atoms of these elements need to gain only 1 electron to fill their outermost energy level. The great reactivity of the halogens explains why they are never found free in nature.

Halogens form compounds in which they share or gain 1 electron. They react with the alkali metals (Family 1) quite easily. One common compound formed when the alkali metal sodium gives up 1 electron to the halogen chlorine is called sodium chloride. You know this compound better as table salt. When halogens react with metals, they form compounds called salts. Perhaps you have heard of sodium fluoride, which is the salt used to fluoridate water and toothpaste, or of calcium chloride, which is used to melt snow and ice on streets and sidewalks. Silver bromide, another halogen salt, is used in photographic film.

Fluorine is the most active halogen. Fluorine and chlorine, which is also highly active, are never found uncombined in nature. As you can see from the periodic table, fluorine and chlorine are gases. Bromine is one of the few liquid elements, while iodine and the metalloid astatine are solids.

The Noble Gases

Your tour across the periodic table ends with Family 18, the **noble gases.** All of the elements in this family are gases that are normally unreactive. But under special conditions, certain noble gases will combine chemically with other elements. Because they do not readily form compounds with other elements, the noble gases are also called the inert gases. The noble, or inert, gases include helium (He), neon (Ne), argon (Ar), krypton (Kr), xenon (Xe), and radon (Rn). Can you figure out why they are so unreactive? Atoms of noble gases already have complete outermost energy levels. They do not need to bond with other atoms. Among the noble gases, helium has 2 valence electrons; neon, argon, krypton, xenon, and radon each has 8 valence electrons.

All of the noble gases are found in small amounts in the Earth's atmosphere. Argon, the most common of the noble gases, makes up only about 1 percent of the atmosphere. Because they are so scarce and so unreactive, the noble gases were not discovered until the end of the nineteenth century. This was almost 50 years after Mendeleev's work.

Some common uses of the noble gases are probably quite familiar to you. No doubt you have seen a balloon filled with helium floating at the end of a string. The brightly colored signs above theaters, restaurants, and stores are filled with inert gases, often called "neon" lights. However, only the red lights are produced by neon. The other colors are produced by argon and several of the other noble gases. Some of the other uses of the noble gases may be less familiar to you. Radon is used to treat certain cancers. Argon and xenon are used in certain light bulbs and lamps.

18

| 2 |
| **He** |
| Helium |
| 4.003 |

| 10 |
| **Ne** |
| Neon |
| 20.179 |

| 18 |
| **Ar** |
| Argon |
| 39.948 |

| 36 |
| **Kr** |
| Krypton |
| 83.80 |

| 54 |
| **Xe** |
| Xenon |
| 131.29 |

| 86 |
| **Rn** |
| Radon |
| (222) |

Figure 5–27
Family 18 is known as the noble gases.

Figure 5–28 *Crystals of xenon tetrafluoride were first prepared in 1962. Before that time, it was believed that noble gases could not form compounds.*

Homely Halogens

Fluorine, chlorine, bromine, and iodine are halogens found in many household substances. Investigate their uses by locating various substances in your environment that contain halogens. Make a chart of your findings. Include examples of the substances if possible.

Rare-Earth Elements

Even though you have completed your tour across the periodic table, you probably have at least one question lurking in the back of your mind. Why are there two rows of elements standing alone at the bottom of the periodic table? The elements in these two rows are called the **rare-earth** elements. The rare-earth elements have properties that are similar to one another. They have been separated out and displayed under the main table to make the table shorter and easier to read.

The first row, called the **lanthanoid series,** is made up of soft, malleable metals that have a high luster and conductivity. The lanthanoids are used in industry to make various alloys and high-quality glass.

The elements in the second row make up the **actinoid series.** All the actinoids are radioactive. (Changes in the nucleus of radioactive atoms cause particles and energy to be given off.) With the exception of the first three elements, all the actinoids are synthetic, or made in the laboratory. The best known actinoid is uranium, which is used as a fuel in nuclear-powered electric generators. You won't be surprised to learn that elements 104 to 109 in period 7 are also synthetic and radioactive.

5–3 Section Review

1. What is the key to the placement of an element in the periodic table?
2. Which two families contain the most active metals?
3. To which family do the most active nonmetals belong?
4. Why are the elements in Family 18 called inert gases?

Critical Thinking—*Applying Concepts*
5. How can the arrangement of elements in the periodic table be used to predict how they will react with other elements to form compounds?

Solving

Completing the Squares

Look at the data for the five elements below. Fill in the missing data. Then construct a square for each element as it would appear in a modern periodic table. In some cases you will have to perform calculations to fill in the missing data. Do not use a completed periodic table in this activity. It is much more fun to do the calculations yourself. How would you use this information in a laboratory?

Data

■ Carbon (symbol ?); atomic mass, 12.01; number of electrons, 6; (atomic number ?)

■ More than 70 percent of the air (name ?); (symbol ?); atomic number, 7; atomic mass, 14

■ (Name ?); O; (number of protons ?); (atomic mass ?); (number of electrons ?)

■ Fluorine (atomic symbol ?); number of protons, 9; atomic mass, 18.99; (atomic number ?)

■ Inert "sign" gas (name ?); (atomic symbol ?); number of electrons, 10; atomic mass, 20.17

atomic number
symbol
name
atomic mass

5–4 Periodic Properties of the Elements

You have learned several ways in which the periodic table provides important information about the elements. Elements in the same family, or vertical column, have similar properties. Elements at the left of the table are metals. Elements at the right are nonmetals. Metalloids, which show properties of both metals and nonmetals, are located on either side of the dark zigzag line.

Additional information about the elements can be obtained from their location in a period, or

Guide for Reading

Focus on this question as you read.

▶ *What periodic trends can be identified in the elements in the periodic table?*

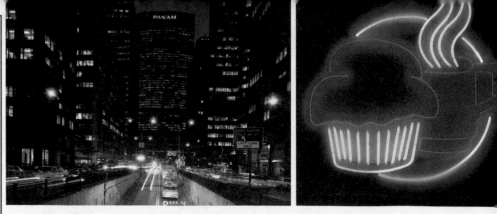

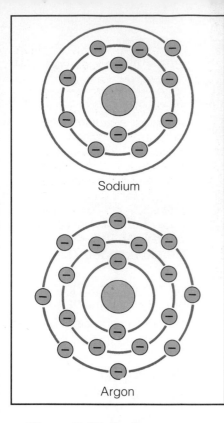

Figure 5–29 *Sodium, an extremely reactive metal, is used in its vapor phase in street lights. Sodium vapor lights provide bright-yellow light (top center). Argon, an extremely unreactive gas, is used to make glowing works of art (top right). How does the electron arrangement of each element account for its reactivity?*

horizontal row. **Certain properties of elements vary in regular ways from left to right across a period. These properties include electron arrangement, reactivity, atomic size, and metallic properties.**

The valence number of an element is related to the electrons in the outermost energy level of an atom of that element. It is these electrons that are involved in the chemical combining of elements to form compounds.

Remember the pattern of valence numbers discovered by Mendeleev: Starting at the left of each period, the pattern of valence numbers is 1 2 3 4 3 2 1 0. An element with a valence of 1 will gain, lose, or share 1 electron in a chemical combination. An element with a valence of 4 will gain, lose, or share 4 electrons. What will happen if an element has a valence of zero? How reactive is such an element?

Elements at the left in a period tend to lose electrons easily when they combine with other elements. You know that the elements at the left of the table are metals. So an important property of metals is that they lose electrons in a chemical combination. Elements at the right in a period tend to gain electrons easily when they combine with other elements. What kinds of elements are these?

The amount of energy needed to remove an electron from an atom shows a periodic increase from left to right across a period. Since atoms of elements at the left in a period tend to lose electrons, removing an electron from such an atom requires a small amount of energy. Removing an electron from an element at the right in a period requires a great amount of energy. Why?

Another property of elements that varies periodically is atomic size. From left to right across a period, atomic size tends to decrease. The decrease can

Figure 5–30 *Fireworks! Colors and noises that astound and amaze! Such brilliant bursts are the results of explosive chemical reactions. Not all chemical reactions are explosive, however. But all chemical reactions do depend on the properties of the elements involved—properties that show a characteristic periodicity.*

be explained in terms of electron arrangement. As the atomic number increases across a period, 1 electron is added to each successive element. But this electron is still in the same energy level. The increase in the number of electrons in the energy level and the number of protons in the nucleus produces a stronger attraction between these oppositely charged particles. The electrons are pulled closer to the nucleus. The size of the atom decreases. Now explain why atomic size increases from top to bottom in a family.

Metallic properties of the elements are also periodic. From left to right across a period, elements become less metallic in nature.

5–4 Section Review

1. What properties are periodic in nature?
2. What is the pattern of valences as you move from left to right across a period?
3. How does atomic size change across a period?

Critical Thinking—*Relating Concepts*
4. A scientist claims that she has discovered a new element that should be inserted between nitrogen and oxygen in the periodic table. Why is it most likely that this scientist has made a mistake? (*Hint:* How does the number of electrons change as you move from left to right across a period?)

CONNECTIONS

The Chains That Bind

Fly from New York to Japan and after hours in the air over land and water, you become convinced of the enormous size of Planet Earth. But in many ways—some more obvious than others—humans are becoming aware of the incredibly small size of the Earth. Although this may seem like a contradiction, in some important ways it is a significant—and alarming—fact.

By now everyone is aware of the dangers of polluting the *environment*. But in the not-too-distant past, such was not the case. The economic growth of major industrial countries was based in large part on the erroneous belief that the enormous size of the Earth made it permissible to dump hazardous substances into the environment. Today we know that, in terms of pollution, the Earth is small. The effects of dumping some hazardous chemicals reach far from their point of introduction into the environment. You are probably familiar with the element mercury. Mercury is an important part of

many industrial processes. For example, it is used in the manufacture of paper and paints. Industrial processes often need a good source of water—especially for washing away the waste byproducts of the manufacturing process. So for a long time, waste mercury was dumped into rivers and streams. After all, the logic was, the water would carry the mercury far away until it eventually reached the vastness of the oceans. There it would be diluted to levels that were no longer dangerous to life.

One major flaw in this logic became evident over time, however. Mercury was found in ever-increasing amounts in the body tissues of certain fishes and other animals. How

fishes store the mercury in their tissues, only in slightly larger amounts. The storage of these chemicals in ever-increasing amounts in the living tissues of organisms in a food chain is called *biological magnification*. By the time mercury has moved through a food chain to reach tuna, birds, cattle, and other larger animals, the amount of stored mercury may have reached levels high enough to threaten health and even endanger survival.

When it comes to certain things—the pollution of water by mercury, for example—the Earth is actually a small place. And the food chains that link one organism to another—the ties that bind—may hold so tightly that organisms at one link of the chain cannot survive.

could the level of mercury reach dangerous proportions in certain organisms? The answer was found by biologists when they examined food chains. A food chain describes a series of events in which food, and therefore energy, is transferred from one organism to another. The first organisms in a food chain are small organisms that are able to produce their own food using simple substances. These organisms are often microscopic and make food by using the energy of the sun or the energy stored in chemicals. Some of these organisms also ingest small amounts of chemicals such as mercury, which is stored in their bodies. In turn, these small organisms are eaten by larger ones—small fishes, for example. And these small

Laboratory Investigation

Flame Tests

Problem

How can elements be identified by using a flame test?

Materials (per group)

nichrome or platinum wire
cork
Bunsen burner
hydrochloric acid (dilute)
distilled water
8 test tubes
test tube rack
8 chloride test solutions
safety goggles

Procedure 🧪 🔥 🖐 ⊡ 👁

1. Label each of the test tubes with one of the following compounds: LiCl, $CaCl_2$, KCl, $CuCl_2$, $SrCl_2$, NaCl, $BaCl_2$, unknown.

2. Pour 5 mL of each test solution in the correctly labeled test tube. Be sure to put the correct solution in each labeled test tube.

3. Push one end of a piece of nichrome or platinum wire into a cork. Then bend the other end of the wire into a tiny loop.

4. Put on your safety goggles. Clean the wire by dipping it into the dilute hydrochloric acid and then into the distilled water. You must clean the wire after you make each test. Holding the cork, heat the wire in the blue flame of the Bunsen burner until the wire glows and no longer colors the burner flame.

5. Dip the clean wire into the first test solution. Hold the wire at the tip of the inner cone of the burner flame. Record the color given to the flame in a data table similar to the one shown here.

6. Clean the wire by repeating step 4.

7. Repeat step 5 for the other six known test solutions. Remember to clean the wire after you test each solution.

8. Obtain an unknown solution from your teacher. After you clean the wire, repeat the flame test for this compound.

Compound		Color of Flame
Lithium chloride	LiCl	
Calcium chloride	$CaCl_2$	

Observations

1. What flame colors are produced by each compound?
2. What flame color is produced by the unknown compound?

Analysis and Conclusions

1. Is the flame test a test for the metal or for the chloride in each compound? Explain your answer.
2. Why is it necessary to clean the wire before you test each solution?
3. What metal is present in the unknown solution? How do you know?
4. How can you use a flame test to identify a metal?
5. What do you think would happen if the unknown substance contained a mixture of two compounds? Could each metal be identified?
6. **On Your Own** Suppose you are working in a police crime laboratory and are trying to identify a poison that was used in a crime. How could a knowledge of flame tests help you?

Study Guide

Summarizing Key Concepts

5–1 Arranging the Elements

▲ The elements in Mendeleev's periodic table are arranged in order of increasing atomic mass.

▲ Mendeleev discovered that the properties of the elements recurred at regular intervals.

▲ Mendeleev left spaces for elements not yet discovered and predicted the properties of these missing elements based on their position in the periodic table.

▲ The modern periodic table is based on the periodic law, which states that the physical and chemical properties of the elements are periodic functions of their atomic numbers.

5–2 Design of the Periodic Table

▲ Horizontal rows of elements are called periods.

▲ Vertical columns of elements are called groups or families.

▲ Elements in the same family have similar properties.

▲ Each square in the periodic table gives the element's name, chemical symbol, atomic number, and atomic mass.

▲ According to their properties, the elements are classified as metals, nonmetals, and metalloids.

5–3 Chemical Families

▲ The number of valence electrons in an atom of an element is the key to its placement in a family in the periodic table.

5–4 Periodic Properties of the Elements

▲ Periodic properties of the elements include electron arrangement, reactivity, atomic size, and metallic properties.

Reviewing Key Terms

Define each term in a complete sentence.

5–1 Arranging the Elements
   periodic law

5–2 Design of the Periodic Table
   group
   family
   period
   metal
   luster
   ductile
   malleable
   corrosion
   nonmetal
   metalloid

5–3 Chemical Families
   alkali metal
   alkaline earth metal
   transition metal
   boron family
   carbon family
   nitrogen family
   oxygen family
   halogen family
   noble gas
   rare-earth element
   lanthanoid series
   actinoid series

Chapter Review

Content Review

Multiple Choice

Choose the letter of the answer that best completes each statement.

1. The periodic law states that the properties of elements are periodic functions of their
   a. mass.
   c. atomic number.
   b. symbol.
   d. valence.
2. Which of the following is a noble gas?
   a. sodium
   c. chlorine
   b. gold
   d. neon
3. When a metal combines with a halogen, the kind of compound formed is called a (an)
   a. organic compound.
   c. actinoid.
   b. salt.
   d. oxide.
4. If a metal can be hammered or rolled into thin sheets the metal is said to be
   a. ductile.
   c. brittle.
   b. malleable.
   d. active.
5. In the periodic table, the metallic character of the elements increases as you move in a period from
   a. right to left.
   c. left to right.
   b. top to bottom.
   d. bottom to top.
6. Which of the following is a halogen?
   a. sodium
   c. carbon
   b. silver
   d. iodine
7. Moseley was able to determine each element's
   a. atomic mass.
   c. symbol.
   b. atomic number.
   d. brittleness.
8. Each period of the table begins on the left with a
   a. highly active metal.
   b. metalloid.
   c. rare-earth element.
   d. nonmetal.
9. Which element is called the "basis of life"?
   a. neon
   c. sodium
   b. carbon
   d. oxygen
10. A brittle element that is not a good conductor of heat and electricity is
    a. inert.
    c. ductile.
    b. a metal.
    d. a nonmetal.

True or False

If the statement is true, write "true." If it is false, change the underlined word or words to make the statement true.

1. Mendeleev noticed a definite pattern in the valence numbers of the elements.
2. The property of a metal that means it can be drawn into thin wire is called luster.
3. Sodium belongs to the transition metals.
4. Nonmetals are usually poor conductors of electricity.
5. The most striking property of the noble gases is their extreme inactivity.
6. In forming compounds, nonmetals tend to lose electrons.
7. Vertical columns of elements in the periodic table are called periods.

Concept Mapping

Complete the following concept map for Section 5–1. Refer to pages N6–N7 to construct a concept map for the entire chapter.

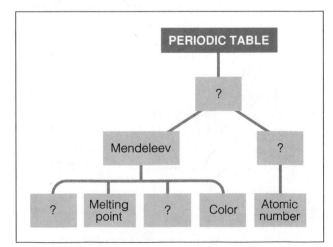

Concept Mastery

Discuss each of the following in a brief paragraph.

1. How was Mendeleev able to tell that there were elements not yet discovered?
2. Compare metals, nonmetals, and metalloids. Give an example of each.
3. Why are elements that gain or lose 1 electron the most active metals or nonmetals?
4. What happens to an atom's size as you move from left to right across a period? From top to bottom in a family?
5. What does the term periodic mean? Give two examples of daily periodic occurrences in your life.

Critical Thinking and Problem Solving

Use the skills you have developed in this chapter to answer each of the following.

1. **Applying definitions** Metals are ductile. In what three ways has this property of metals affected your daily life?
2. **Interpreting charts** Look at these three squares from the periodic table. What kinds of information can you gather from these squares? How would you use this information in a laboratory?

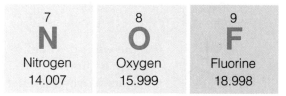

3. **Relating concepts** Use the periodic table to help predict what will happen when the elements in each of the following pairs are brought together in a chemical reaction:
   a. barium and oxygen
   b. lithium and argon
   c. potassium and iodine
   d. sodium and bromine
4. **Applying concepts** Determine the identity of the following elements:
   a. This nonmetal has 4 valence electrons, properties similar to carbon, and an atomic mass slightly less than phosphorus.
   b. This element has 5 valence electrons, shows properties of metals and nonmetals, and has 33 protons in the nucleus of each atom.
   c. This highly active metal is a liquid. It has 1 valence electron.
5. **Making diagrams** Draw a diagram to show the arrangement of electrons in the outermost energy level of an atom in each family in the periodic table.
6. **Classifying elements** Classify each of the following elements as very active, moderately active, fairly inactive, or inert: magnesium, mercury, fluorine, krypton, helium, gold, potassium, calcium, bromine.
7. **Using the writing process** Suppose Mendeleev was alive today and his development of the periodic table occurred yesterday. Plan and write a script for a "media event" that would bring Mendeleev's work to the attention of the public. For example, you might want to produce a television interview with Mendeleev. Or you might want to write a newspaper article.

SCIENCE GAZETTE

SHIRLEY ANN JACKSON: Helping Others Through SCIENCE

▲ Shirley Ann Jackson, in her office at Bell Laboratories, is presently doing research in the field of optoelectronic materials used in communication devices.

 magine what it would be like to catch a glimpse of the universe as it was forming—to look back in time nearly 20 billion years! Of course, no one can really see the beginning of time. But physicists such as Shirley Ann Jackson believe that learning about the universe as it was in the past will help us understand the universe as it is now and as it will be in the future.

By unraveling some of the mysteries of the universe, Dr. Jackson hopes to fulfill a basic ambition: to enrich the lives of others and to make the world a better place in which to live. This contribution, Dr. Jackson believes, can be achieved through science.

Jackson was born and raised in Washington, DC. After graduating from high school as valedictorian, she attended Massachusetts Institute of Technology, MIT. There, her role as a leader in physics began to take root. Jackson became the first

African-American woman to earn a doctorate degree from MIT. She is also the first African-American woman to earn a PhD in physics in the United States.

After graduate school, Jackson began work as a research associate in high-energy physics at the Fermi National Accelerator Laboratory in Batavia, Illinois. This branch of physics studies the characteristics of subatomic particles—such as protons and electrons—as they interact at high energies.

Using devices at Fermilab called particle accelerators, physicists accelerate subatomic particles to speeds that approach the speed of light. The particles collide and produce new subatomic particles. By analyzing these subatomic particles, physicists are able to learn more about the structure of atoms and the nature of matter.

The experiments in which Jackson participated at Fermilab helped to prove the existence of certain subatomic particles whose identity had only been theorized. This information is important in understanding the nuclear reactions that are taking place at the center of the sun and other stars.

Jackson's research is not limited to the world of subatomic particles alone. Her work also includes the study of semiconductors—materials that conduct electricity better than insulators but not as well as metal conductors. Semiconductors have made possible the development of transistor radios, televisions, and computers—inventions that have dramatically changed the ways we live.

Jackson's current work in physics at Bell Laboratories in Murray Hill, New Jersey, has brought her from the beginnings of the universe to the future of communication. This talented physicist has been doing research in the area of optoelectronic materials. This branch of electronics—which deals with solid-state devices that produce, regulate, transmit, and detect electromagnetic radiation—is changing the way telephones, computers, radios, and televisions are made and used.

Looking back on her past, Jackson feels fortunate to have been given so many opportunities at such a young age. And she is optimistic about the future. "Research is exciting," she says. Motivated by her research, Shirley Ann Jackson is happy to be performing a service to the public in the way she knows best—as a dedicated and determined scientist.

▼ **This particle-accelerator generator at Fermilab is familiar equipment to Shirley Ann Jackson.**

ACID RAIN:
It Won't Go Away

▲ Although many lakes look clean and beautiful, their waters will no longer support life. Unable to live in highly acidic water, dead fish wash up on shore.

High in the Adirondack Mountains of northern New York State, a crystal-blue lake shimmers in the sun. Cradled by green slopes, the lake once held huge trout, which drew anglers from all over the country. But now the lake has no fish. Why? For years the lake has been pelted by polluted rain that is almost as acid as vinegar.

◀ **Tall smokestacks spew clouds of chemicals high into the air. The chemicals can mix with rain, resulting in acid rain.**

Unable to survive the high acid content of the water, the fish died.

The tragedy of acid rain has not struck just this one lake in the Adirondacks. At least 180 trout ponds and lakes in these mountains no longer support fish, according to the National Wildlife Federation. What is more, acid rain is not confined to the Adirondacks.

Acid rain has fallen from coast to coast. Rainfall in general seems to have become more acidic. This has led conservation groups such as the National Wildlife Federation to issue warnings about acid rain's danger to the environment. "Acid rain is a growing threat," says the federation. The wildlife federation estimates that "acid rain may be slowly poisoning 160,000 kilometers of streams and 20,000 lakes" across the United States, mostly in the eastern half of the country. Acid rain also kills trees and wears down bricks, concrete, and statues made of stone and metal. It may even affect our drinking water.

HOW BAD IS IT?

Acid rain is a fact. But not all scientists agree on how widespread or how dangerous it is. "There is no question that the northeastern United States is experiencing acid rains,"

admits F. D. Bess, an environmental engineer for the Union Carbide Corporation. But he adds that the amount of increase in rain's acidity and the regions it affects are uncertain. And the American Chemical Society has said that earlier measurements of the amount of acid in rain may have been wrong. So it may be difficult to compare the acidity of rain today with, for instance, rain in the 1950s.

Some biologists representing fishing groups say that the danger may not be as great as many conservationists believe. "We are concerned but not going as far out as other environmental groups," says Bob Martin, a biologist for the Sports Fishing Institute. And scientists who work for industries blamed for acid rain argue that before industry can be condemned, many more facts about acid rain must be learned.

Conservationists and industry scientists alike agree that much about acid rain is a mystery. Acid rain (the term includes snow, fog, and other moisture from above) is formed from two types of chemical pollutants: sulfur dioxide and oxides of nitrogen. These pollutants react with moisture in the atmosphere to form sulfuric acid and nitric acid. Then when the moisture falls, it is acidic.

WHERE DOES IT COME FROM?

The chemicals that cause acid rain can also enter the atmosphere naturally—from volcanoes and forest fires, for instance. According to the American Chemical Society, "natural emissions of sulfur into the atmosphere are thought to be greater than industrial sources." Actually, "pure" rain is naturally a bit acidic. But pollutants also come from automobile engines and from industrial factories that burn coal and other fossil fuels. These factories, in the eyes of many conservationists, are the real culprits.

One of the most important sources of acid rain in the Northeast, according to the

National Wildlife Federation and other conservation groups, is the Ohio Valley industries. Especially cited are those industries with smokestacks that reach high into the air. Conservationists claim that winds blowing east from the Ohio Valley carry pollutants to places such as the Adirondacks, where they fall as acid rain. But many scientists who have studied the problem say that tracing acid rain to its source is still difficult.

Kenneth A. Rahn, a University of Rhode Island scientist, has been developing a method to trace chemical pollutants. He has found, for example, that some of the acid rain falling on the Adirondack lakes may come from Canada and not from the Ohio Valley. Meanwhile, Canadians have claimed that acid rain from American industries is polluting their lakes. The Canadians have demanded that the United States government do something about it.

Spokespeople for the Ohio Valley region have also blamed other sources for the problem. In 1980, an environmental official from Pennsylvania, William B. Middendorf, told Congress that the acid rain falling on the parts of the Ohio Valley that are in his state may be caused by pollutants from the Midwest. If this is true, then the Ohio Valley may not be the only major source of the rain that is killing Adirondack trout.

IS THERE A SOLUTION?

The effects of acid rain depend on the location of the lake. Even if acid rain is a widespread problem, not all lakes are affected by it in the same way. Many Adirondack lakes lie on steep slopes and are surrounded by thin soil. Rainwater passes through the soil quickly. Minerals in the soil, such as limestone, that could remove acids do not have time to work. Some Adirondack lakes lie on granite, which does not permit water to seep through into the soil. So the acids build up in the lakes. In lakes lying on limestone, however, acids are neutralized.

Ironically, other types of pollution can work against the effects of acid rain. Sewage,

▲ **Acid rain eats away at stone and metal statues and monuments. Acid rain is probably one of the factors that has caused this statue to decay.**

for example, often contains chemicals that undo the effects of acids. So lakes that are in the wilderness away from people may be more threatened by acid rain than lakes in cities.

In the long run, a solution must be found to the problem of acid rain. The National Academy of Sciences suggests that industry reduce by 12 million tons the amount of chemical pollutants being released into the air. One way to do this is to install devices called "scrubbers" in smokestacks. The scrubbers clean emissions before they are released into the air. Another solution is to use low-sulfur coal, which creates less pollution than other types of coal. But that would be expensive and would possibly cost some coal miners their jobs.

Someone will have to pay for solving the acid rain problem. Should it be industry? Should the government—that is, the taxpayers—pay the bill? If the pollution does come from the Ohio Valley, should that part of the country pay? Or should the Northeast help? Acid rain and the questions it raises will be around for a long time.

FACTORIES BEYOND EARTH

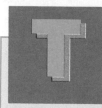

The sun's intense rays bounced off giant mirrors and were instantly focused on the lump of iron hanging in space. As the temperature of the iron rose, it began to melt. Quickly, two white-suited astronauts floating out of range of the hot rays added a little carbon and aluminum to the molten iron.

"Maybe we should add some nickel from that last asteroid the space miners brought back," one of the astronauts radioed to her co-worker. "We want that steel we're making to be tough and acid-resistant."

"I'll call the customer on my long-range radio and let you know," the other astronaut replied. "Just let the steel hang there in a molten lump until I get the answer."

Meanwhile, inside a nearby space factory, a chemist was busy growing crystals for electronic circuits. "I wonder what it was like to grow crystals like these on Earth,

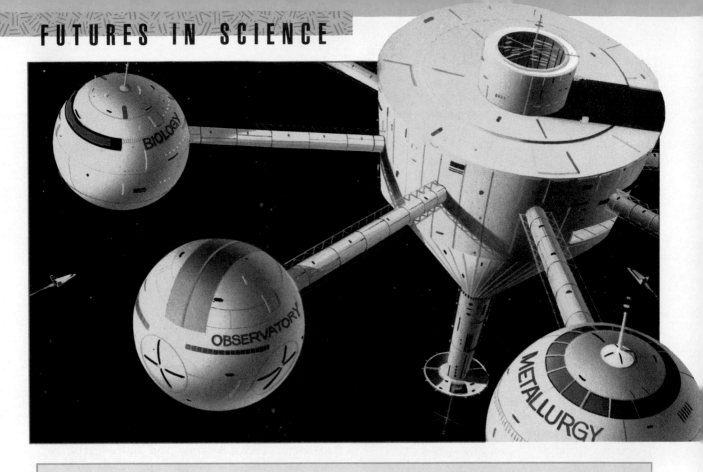

where gravity kept them from forming the perfect shape they form here in space," he thought to himself. "I guess chemists had problems in the days when there were no gravity-free laboratories or processing plants. The crystals I'm growing will be used to make the very best computers in the solar system."

Down the hall from the chemist, a biology professor was telling her medical students about the advantages of making medicines in space. "We know from an experiment performed in space way back in 1975 that kidney cells produce much more of a special chemical when they are grown in space than when they are grown on Earth," she explained. "We have recently discovered that, like kidney cells, human pancreas cells will produce more of their special chemical, insulin, when they are grown in space labs. We are also growing bacteria in space that can produce human insulin. Bacteria grow faster in space than on Earth. Growing more human insulin at a faster rate is a big help to people who have diabetes.

"We're also making purer vaccines and medicines here in space than we ever made on Earth," the professor continued.

"Why is that?" asked a student.

"Because we don't need to use containers to hold the materials we're mixing," she responded. "Without gravity, they hold together all by themselves. And because we can eliminate containers, we avoid contaminating the materials we're working with. On Earth, microscopic pieces of containers unavoidably got mixed in with the products. There was no such thing as a really pure product on Earth—whether it was a medicine, cosmetic, metal, or glass."

SPACE METALS

In another space factory a few thousand kilometers away, astronauts were setting up a seemingly strange experiment with metals. "Just let those metals sit outside on the dark side of the factory for a few minutes. They will cool off quickly in dark space because they are shielded from the sun there. Then we'll see what happens to the metals as they get colder and colder," the chief scientist told the astronauts. "We want to see how these new mixtures of metals, or alloys, stand up to the cold of dark space. Will they stay tough yet flexible out in space, or will they become brittle and shatter if something hits them? Will dust-sized micrometeorites damage them? Or would the object have to be baseball-sized, car-sized, or even larger? Which metals will 'survive' when they are half in sunlight and half in shadow?"

"These alloys could never have been made on Earth," observed one of the astronauts.

"That's right," responded the scientist. "They can be produced only in zero gravity. But there are many questions to be answered about these metals before we can use them in building space colonies."

In another part of the factory, workers were making superlight, high-strength alloys for use both on Earth and in space. Some of these metals would be used as thin, protective coatings on everything from space shuttles to earthbound craft.

HOLD THAT VIRUS!

In a space laboratory beyond the moon, medical researchers were performing experiments in genetic engineering. They were trying to find cures for age-old diseases by attempt-ing to "redesign" viruses and bacteria.

"We would never have been able to do this kind of research on Earth," one of the older scientists said to a new lab assistant. "It's much too risky. On Earth, if some of the altered bacteria or viruses 'escaped' from the lab, we might be in big trouble. And if we didn't know how to control them, we might find ourselves in the middle of a major epidemic. Here a million and a half kilometers from Earth, we can work with more safety controls than we ever could have on Earth. Our research can go on without endangering people in other space colonies or on Earth."

DOWN-TO-EARTH QUESTIONS

What else will we be able to make in the airlessness of space? In zero gravity? In the heat of direct, focused sunlight? In the supercold regions shaded from the sun? What will we build with our perfectly shaped crystals, our pure chemicals, and our metal alloys that cannot be produced on Earth? What medical breakthroughs will we make in space? What dangerous but vital research can we perform in space while Earth is protected from its risks?

Will you be one of the space researchers who will answer these questions?

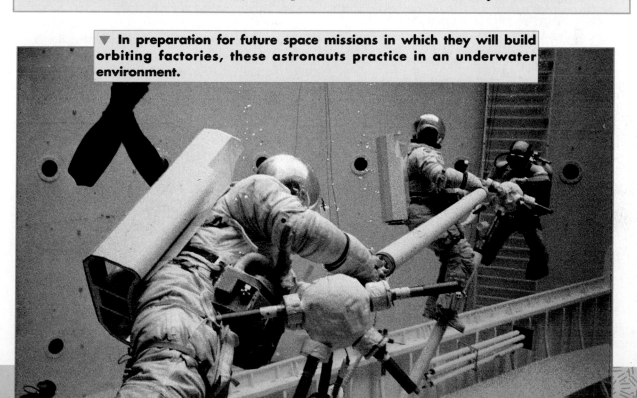

▼ **In preparation for future space missions in which they will build orbiting factories, these astronauts practice in an underwater environment.**

For Further Reading

> If you have been intrigued by the concepts examined in this textbook, you may also be interested in the ways fellow thinkers—novelists, poets, essayists, as well as scientists—have imaginatively explored the same ideas.

Chapter 1: General Properties of Matter

Dahl, Roald. *Charlie and the Chocolate Factory*. New York: Penguin.

Lipsyte, Robert. *One Fat Summer*. New York: Bantam.

Verne, Jules. *Around the World in Eighty Days*. New York: Bantam.

————. *Twenty Thousand Leagues Under the Sea*. New York: Pendulum Press.

Chapter 2: Physical and Chemical Changes

Crichton, Michael. *The Andromeda Strain*. New York: Knopf.

Hawthorne, Nathaniel. "The Birthmark," in *Tales and Sketches*. New York: Viking Press.

McKillip, Patricia. *The Changeling Sea*. New York: Ballantine.

O'Dell, Scott. *Black Pearl*. New York: Dell.

Chapter 3: Mixtures, Elements, and Compounds

Hamilton, Virginia. *Arilla Sun Down*. New York: Greenwillow.

Mahy, Margaret. *The Catalogue of the Universe*. New York: Macmillan.

Plotz, Helen. *Imagination's Other Place: Poems of Science and Math*. New York: Crowell.

Shelley, Mary Wollstonecraft. *Frankenstein*. New York: Bantam.

Chapter 4: Atoms: Building Blocks of Matter

Asimov, Isaac. *How Did We Find Out About Atoms?* New York: Walker & Co.

Feynman, Richard. *Surely You're Joking, Mr. Feynman! Adventures of a Curious Character*. New York: Bantam.

Larsen, Rebecca. *Oppenheimer and the Atomic Bomb*. New York: Watts.

Keller, Mollie. *Marie Curie*. New York: Watts.

Chapter 5: Classification of Elements: The Periodic Table

Konigsburg, E. L. *Father's Arcane Daughter*. New York: Atheneum.

Levi, Primo. *The Periodic Table*. New York: Schocken.

Lowell, Amy. "Patterns" in *Selected Poems of Amy Lowell*. Cambridge. MA: Riverside Press.

Paton Walsh, Jill. *A Parcel of Patterns*. New York: Farrar, Straus & Giroux, Inc.

Activity Bank

Welcome to the Activity Bank! This is an exciting and enjoyable part of your science textbook. By using the Activity Bank you will have the chance to make a variety of interesting and different observations about science. The best thing about the Activity Bank is that you and your classmates will become the detectives, and as with any investigation you will have to sort through information to find the truth. There will be many twists and turns along the way, some surprises and disappointments too. So always remember to keep an open mind, ask lots of questions, and have fun learning about science.

WHAT "EGGS-ACTLY" IS GOING ON HERE?

This simple investigation can help you observe and better understand the concept of density and the effects of different densities. You might like to share the results of this investigation with your family and friends. It may amaze and astound them.

You Will Need

hard-boiled egg
saucer
large clear plastic
   container

teaspoon
box of table salt

Now You Can Begin

1. You might be able to buy a hard-boiled egg from a delicatessen or your school's cafeteria. If you cannot, it won't take long to cook a hard-boiled egg yourself. *Ask a parent or other adult to help you with this procedure.* Place the egg in a small pan filled with cool water. Put the pan on a heat source. When the water starts to boil, begin to count off 10 minutes. At the end of this time, the egg is properly cooked. *Again ask an adult to help you now. Use a large*

spoon to remove the egg from the water. Place the egg in a small saucer and let it cool. As an alternative, you can also place the pan in the sink and let cold water run into the pan. The egg will cool more quickly this way.

2. Carefully place the egg at the bottom of your plastic container. Add enough water so that the container is three-quarters full. The egg should be covered by at least 10 cm of water.

3. Describe what you see and make a drawing of your setup.

4. Gradually add salt to the water in your container. Add the salt 1 teaspoon at a time. Stir the water carefully. Do not break the egg. If you are careful you can make egg salad at the end of this investigation! Describe what happens after you add each teaspoon of salt.

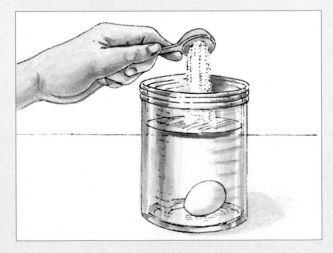

Seeing and Explaining

1. The position of the egg changed at some point. Describe any changes that occurred after adding salt.

2. Why do you think the changes occurred?

A Different View

Your family is undecided about where to spend summer vacation. Traditionally the family takes a vote, with each member casting one vote. The choice is between a cabin at a mountain lake or a trip to the seashore. Your younger brother likes to swim, although he does not swim very well. He will, however, cast the deciding vote. Be a "good egg" and apply what you have learned in this investigation to help your brother make a wise choice. Outline your thinking here.

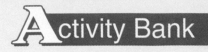
CRYSTAL GARDENING

Diamonds, sugar, and table salt have something in common. They all form crystals. In fact, many substances form crystals, but usually only under certain conditions. In this investigation you will be able to "grow" some salt or sugar crystals in the laboratory. You probably would also like to grow diamonds, but unfortunately the conditions needed to produce diamond crystals cannot be duplicated in the laboratory! You can work with a partner in this investigation.

Materials Needed

table sugar
table salt
2 small metal washers or nuts
2 wooden tongue depressors or sticks
    from an ice cream bar
2 heat-proof plastic jars
sauce pan
source of heat
teaspoon
string

What to Do

1. Pour a cup of water into a sauce pan and place the pan on a source of heat. **CAUTION:** *You will need an adult's permission and help during this part of the investigation.*

2. Add 1 teaspoon of salt to the water as it heats. Do not let the water boil. Keep stirring the solution. Keep adding salt, 1 teaspoon at a time. Observe the water as you add the salt. At some point, after you have added a quantity of salt, any additional salt will appear to take a long time to dissolve. At this point, add 1 more teaspoon of salt. Remove the pan from the heat.

3. Put the pan on a heat-resistant surface. Prepare the rest of your apparatus as the water cools. Tie one end of a piece of string to the washer or nut. Wind the other end of the string around the piece of wood as shown in the diagram. Adjust the length of the string so that when the piece of wood is placed across the top of the jar, the washer at the other end of the string does not touch the bottom of the jar.

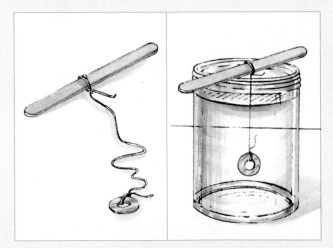

4. When the pan has cooled, carefully pour the saltwater solution into the jar. Place the jar in a spot where it can remain undisturbed for several days.

5. Repeat the above procedure using table sugar instead of table salt.

Observations

1. After several days examine the two jars. Describe what you see. Accompany your description with a drawing.

2. What was the solvent and the solute in each of the solutions from which you grew your crystal gardens?

Something to Think About

On June 25, 1992, NASA launched space shuttle Columbia on its longest mission yet. One of the experiments that was conducted on this mission was to grow crystals of zeolite, a compound formed from aluminum and silica. Scientists wanted to see if larger and more perfect crystals of this compound could be produced in space. What variable do you think scientists were testing in this experiment? Explain your answer.

HOW TO WATCH THE FOODS YOU EAT

Today almost everyone is watching their weight and is showing concern about the foods they eat. To be a wise consumer you should know what nutrients are contained in the foods you eat. This investigation will show you ways in which you can test the foods you eat to see if they contain fats and starches.

Materials

small samples of cooking oil, butter, bread, cut piece of potato, carrot, cooked bacon, potato or corn chip, other food samples

iodine solution

2 medicine droppers

large brown paper bag

scissors

Before You Start

You will use two simple tests to find out if your food samples contain fats and starches. You can use the results of the first two tests you perform as comparisons for your other samples.

1. Use a medicine dropper to place a drop of the iodine solution on the cut side of a piece of potato. **CAUTION:** *Be careful not to spill the iodine solution on your clothing. It will stain.* A potato con-

tains starch. The reaction you observe when iodine is placed on a potato is considered a positive test for starch. Record your observations in a data table.

2. Test other food samples with the iodine solution. Compare these results with the potato-iodine test you performed in step 1. Record these results in your data table.

3. Use your scissors to cut the brown paper bag along the sides. Open up the bag to form a single sheet of paper. Use a clean medicine dropper to place a drop of cooking oil on an area of the paper bag. Hold the paper up to a source of light and observe the spot. Record your observations in your data table. The spot you observe is considered to be a positive test for the presence of fats.

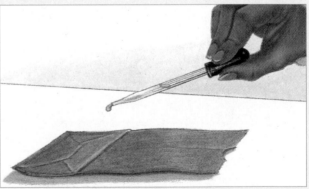

4. Test other food samples by rubbing a small amount of each sample on brown paper if it is a solid or by placing a drop of the sample on the brown paper if it is a liquid. Compare the results to the result you obtained with the cooking oil. Record your observations in your data table.

Observations

1. What is a test for starch?
2. What foods contain starch?
3. What is a test for fats?
4. What foods contain fats?

Using What You Learned

1. Did the results of any of these tests surprise you? Why?
2. These two tests are qualitative tests; that is, they show whether a food contains fats or starches. However, they do not show how much fat or starch a food contains. Why would it be important to know how much fat or starch is in a particular food?

On Your Own

Prepared foods list the amounts of nutrients they contain on their label. Collect some food labels and analyze them. Prepare a report for your classmates about your findings.

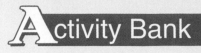
ACID RAIN TAKES TOLL ON ART

You may have read headlines similar to the one above and wondered how a gentle rain, pitter-pattering on your window panes, could be strong enough to wear away stone. Imagine that a city government has hired you to investigate how acid rain and other forms of acid precipitation are affecting its buildings and outdoor statues. How would you go about this task? What problems would you encounter? What solutions would you propose? Building a model in your laboratory might help you to study the problem of acid rain.

You Will Need

piece of soft chalk	shallow pan
nail	petri dish
clear vinegar	plastic wrap
medicine dropper	

Now to Work 👁

1. Place the shallow pan in front of you on your desk or table. Working over the shallow pan will make cleanup easier when you are finished. Use the nail to carve a figure from the piece of chalk. (Chalk is calcium carbonate, the compound that makes up limestone and marble.) Include some surface details in your carving—eyes and hair are

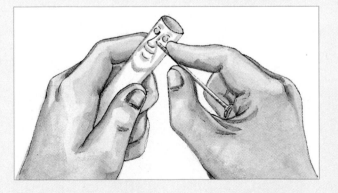

good. Remember, you will not be graded on your artistic abilities!

2. Place your carving in the middle of the pan. Fill a medicine dropper with vinegar. Let several drops of the vinegar drip onto your "work of art." Vinegar contains acetic acid diluted in water.

3. Repeat the vinegar treatment five more times. Observe the effects each time.

4. Place a petri dish in the pan and set your carving on its side in the dish. Pour enough vinegar into the petri dish to just cover your carving.

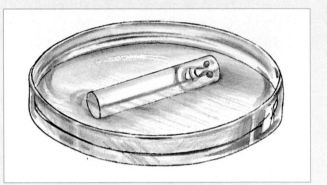

5. Cover the petri dish with plastic wrap and store it overnight in a place where it will not be disturbed.

6. Observe your carving after the overnight treatment.

Observing and Thinking

1. What happened to your carving when vinegar was dropped on it?

2. How did the length of time your carving was exposed to vinegar affect the details on your carving? How do your findings relate to city buildings and sculptures?

3. Studies have shown that the rate at which acid rain dissolves limestone and

marble is much slower than the rate at which vinegar dissolves chalk. How would this fact affect your study?

4. What steps could be taken to prevent further damage to city buildings and sculptures from acid rain? Be innovative in your suggestions.

5. What long-term action should be taken to eliminate the problem of acid rain?

Cooperating With Others

Library research: Find out the causes of acid rain and other forms of acid precipitation. Make a report on your findings. Offer suggestions to deal with this problem.

Field work: With a teacher's or parent's permission, make a survey of the effects of acid precipitation on the buildings or statues in your city or town. If you are fortunate enough to live in an area where acid rain is not a problem, you might like to make a survey of the effects of other environmental conditions on buildings and statues. You might like to make drawings or take photographs to illustrate your findings.

Several friends or classmates can work with you on this project. Their skills, ideas, and abilities can improve the quality of a project.

HUNTING FOR TREASURE IN TRASH

This is a long-term research project you can do with several of your classmates. What you discover should be shared with your class—maybe even with other classes in your school.

Some Help Before You Begin

All substances—the foods you eat, the clothes you wear, the bus that takes you to school, and the spaceship that may some-day take you to the moon—are made of matter. You already know that all matter is made of elements—elements alone or elements combined with other elements to form compounds.

Planet Earth, in fact, can be thought of as a giant "element bank or storehouse" in which elements are withdrawn, used, and eventually discarded or recycled. The ways in which we live our lives determine whether these materials are used wisely.

Located below are several ideas that offer some direction for your research project. Your project should begin in your library and is limited in scope only by your desires. A good book for you to borrow from your local library is: *How to Do Successful Science Projects* by Norman F. Smith (Revised Edition), published by Julian Messner in 1990.

1. Find out which elements used by modern society to make different substances are reusable. Find out if your community is recycling materials. You might like to develop, with your classmates, a program to recycle materials in your school. Check your plan with your teacher.

2. Nature also recycles materials. When an animal or a plant dies, the substances that make it up are used again as part of the continuous cycle of life on Planet Earth. You might like to investigate the rate at which different kinds of materials break down in the environment. A series of experiments can be set up to determine the rate of breakdown, or even if materials will break down. After you develop a plan of study, check with your teacher for his or her ideas about the soundness of your proposed project.

Eventually you might like to make a presentation to your school board or even to the mayor of your town about your findings and suggestions.

WHAT IS THE EFFECT OF PHOSPHATES ON PLANT GROWTH?

One warm, sunny day, Frank and Marie decide to go for a hike in the country. After walking for some time, they see a lake in the distance. It looks beautiful, and from afar seems like an ideal place to stop and rest. As they approach the water, however, they notice an awful smell. It smells as if something is rotting. When they reach the shore, Frank and Marie notice that the lake is covered with a green mat of water plants. They notice dead fish floating. What could have caused this disaster?

Sometimes seemingly harmless chemicals have effects that are not easily predictable. For example detergents are often used to clean clothes and dishes. Phosphates, a group of chemicals found in some detergents, have been banned by some communities. In this investigation you will measure the effects of phosphates on plant growth. Then, maybe, you will have a clue to explain the observations of the two unhappy hikers.

Materials

2 large test tubes with corks or stoppers to fit
test-tube holder, or large plastic jar or beaker
2 sprigs of *Elodea*
detergent that contains phosphates
medicine dropper
sunlight or a lamp
small scissors

Before You Begin

Make sure that the detergent you use contains phosphates. Many do not. *Elodea*

is a common water plant used in home aquariums. A local pet store is a good source of supply.

Procedure

1. Take two sprigs of *Elodea*. Use your scissors to cut the sprigs to the same length. Measure the length of the sprigs and record the length in a data table similar to the table below. Place a sprig of *Elodea* into each test tube.

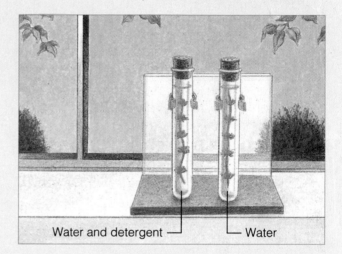
Water and detergent —   — Water

2. Add enough water to each test tube to fill it nearly to the top. Be sure to cover the *Elodea* sprig with water.
3. Place a small pinch of detergent into one test tube. Gently swirl the test tube to mix the water and detergent. Leave plain water in the other test tube.
4. Stopper each test tube.
5. Place the test tubes in a test-tube rack or jar. Place the rack or jar in a sunny window or under another source of light.

(continued)

6. Every three days for a month, carefully remove and measure each *Elodea* sprig. Record your measurements in your data table. Place the sprigs back into the test tubes from which they were removed. **Note:** *Do not mix up the sprigs.*

DATA TABLE

	Day	3	6	9	12	15	18	21	24	27	30
Water											
Water + Detergent											

Observations

1. What was the control in this experiment? Why?

2. Describe the *Elodea* that was placed in plain water.

3. Describe the *Elodea* that was placed in water that contained the detergent drops.

4. Why was it important to return each sprig to the correct test tube?

Analysis and Conclusions

1. Did the detergent affect the *Elodea's* growth?

2. How do you explain the result of this investigation?

Going Further

Most of the communities that ban the use of detergents that contain phosphates are concerned about the quality of their water supply. How might the effect of phosphates on water plants affect a community's water supply?

Appendix A

The metric system of measurement is used by scientists throughout the world. It is based on units of ten. Each unit is ten times larger or ten times smaller than the next unit. The most commonly used units of the metric system are given below. After you have finished reading about the metric system, try to put it to use. How tall are you in metrics? What is your mass? What is your normal body temperature in degrees Celsius?

Commonly Used Metric Units

Length   The distance from one point to another

meter (m)	A meter is slightly longer than a yard.
	1 meter = 1000 millimeters (mm)
	1 meter = 100 centimeters (cm)
	1000 meters = 1 kilometer (km)

Volume   The amount of space an object takes up

liter (L)	A liter is slightly more than a quart.
	1 liter = 1000 milliliters (mL)

Mass   The amount of matter in an object

gram (g)	A gram has a mass equal to about one paper clip.
	1000 grams = 1 kilogram (kg)

Temperature   The measure of hotness or coldness

degrees Celsius (°C)	0°C = freezing point of water
	100°C = boiling point of water

Metric–English Equivalents

2.54 centimeters (cm) = 1 inch (in.)
1 meter (m) = 39.37 inches (in.)
1 kilometer (km) = 0.62 miles (mi)
1 liter (L) = 1.06 quarts (qt)
250 milliliters (mL) = 1 cup (c)
1 kilogram (kg) = 2.2 pounds (lb)
28.3 grams (g) = 1 ounce (oz)
°C = 5/9 × (°F − 32)

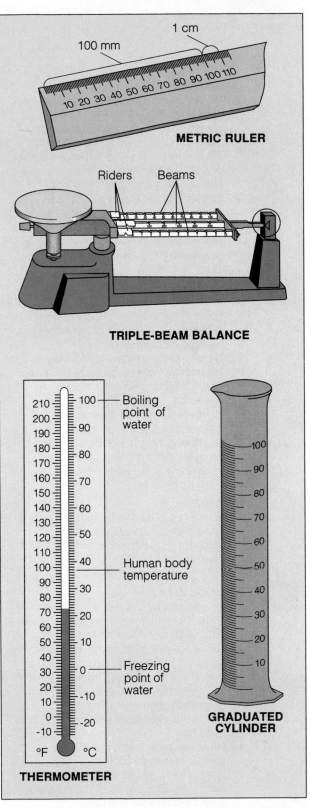

METRIC RULER

TRIPLE-BEAM BALANCE

Riders   Beams

Boiling point of water

Human body temperature

Freezing point of water

THERMOMETER

GRADUATED CYLINDER

Glassware Safety

1. Whenever you see this symbol, you will know that you are working with glassware that can easily be broken. Take particular care to handle such glassware safely. And never use broken or chipped glassware.
2. Never heat glassware that is not thoroughly dry. Never pick up any glassware unless you are sure it is not hot. If it is hot, use heat-resistant gloves.
3. Always clean glassware thoroughly before putting it away.

Fire Safety

1. Whenever you see this symbol, you will know that you are working with fire. Never use any source of fire without wearing safety goggles.
2. Never heat anything—particularly chemicals—unless instructed to do so.
3. Never heat anything in a closed container.
4. Never reach across a flame.
5. Always use a clamp, tongs, or heat-resistant gloves to handle hot objects.
6. Always maintain a clean work area, particularly when using a flame.

Heat Safety

Whenever you see this symbol, you will know that you should put on heat-resistant gloves to avoid burning your hands.

Chemical Safety

1. Whenever you see this symbol, you will know that you are working with chemicals that could be hazardous.
2. Never smell any chemical directly from its container. Always use your hand to waft some of the odors from the top of the container toward your nose—and only when instructed to do so.
3. Never mix chemicals unless instructed to do so.
4. Never touch or taste any chemical unless instructed to do so.
5. Keep all lids closed when chemicals are not in use. Dispose of all chemicals as instructed by your teacher.

6. Immediately rinse with water any chemicals, particularly acids, that get on your skin and clothes. Then notify your teacher.

Eye and Face Safety

1. Whenever you see this symbol, you will know that you are performing an experiment in which you must take precautions to protect your eyes and face by wearing safety goggles.
2. When you are heating a test tube or bottle, always point it away from you and others. Chemicals can splash or boil out of a heated test tube.

Sharp Instrument Safety

1. Whenever you see this symbol, you will know that you are working with a sharp instrument.
2. Always use single-edged razors; double-edged razors are too dangerous.
3. Handle any sharp instrument with extreme care. Never cut any material toward you; always cut away from you.
4. Immediately notify your teacher if your skin is cut.

Electrical Safety

1. Whenever you see this symbol, you will know that you are using electricity in the laboratory.
2. Never use long extension cords to plug in any electrical device. Do not plug too many appliances into one socket or you may overload the socket and cause a fire.
3. Never touch an electrical appliance or outlet with wet hands.

Animal Safety

1. Whenever you see this symbol, you will know that you are working with live animals.
2. Do not cause pain, discomfort, or injury to an animal.
3. Follow your teacher's directions when handling animals. Wash your hands thoroughly after handling animals or their cages.

Appendix C

One of the first things a scientist learns is that working in the laboratory can be an exciting experience. But the laboratory can also be quite dangerous if proper safety rules are not followed at all times. To prepare yourself for a safe year in the laboratory, read over the following safety rules. Then read them a second time. Make sure you understand each rule. If you do not, ask your teacher to explain any rules you are unsure of.

Dress Code

1. Many materials in the laboratory can cause eye injury. To protect yourself from possible injury, wear safety goggles whenever you are working with chemicals, burners, or any substance that might get into your eyes. Never wear contact lenses in the laboratory.

2. Wear a laboratory apron or coat whenever you are working with chemicals or heated substances.

3. Tie back long hair to keep it away from any chemicals, burners and candles, or other laboratory equipment.

4. Remove or tie back any article of clothing or jewelry that can hang down and touch chemicals and flames.

General Safety Rules

5. Read all directions for an experiment several times. Follow the directions exactly as they are written. If you are in doubt about any part of the experiment, ask your teacher for assistance.

6. Never perform activities that are not authorized by your teacher. Obtain permission before "experimenting" on your own.

7. Never handle any equipment unless you have specific permission.

8. Take extreme care not to spill any material in the laboratory. If a spill occurs, immediately ask your teacher about the proper cleanup procedure. Never simply pour chemicals or other substances into the sink or trash container.

9. Never eat in the laboratory.

10. Wash your hands before and after each experiment.

First Aid

11. Immediately report all accidents, no matter how minor, to your teacher.

12. Learn what to do in case of specific accidents, such as getting acid in your eyes or on your skin. (Rinse acids from your body with lots of water.)

13. Become aware of the location of the first-aid kit. But your teacher should administer any required first aid due to injury. Or your teacher may send you to the school nurse or call a physician.

14. Know where and how to report an accident or fire. Find out the location of the fire extinguisher, phone, and fire alarm. Keep a list of important phone numbers—such as the fire department and the school nurse—near the phone. Immediately report any fires to your teacher.

Heating and Fire Safety

15. Again, never use a heat source, such as a candle or burner, without wearing safety goggles.

16. Never heat a chemical you are not instructed to heat. A chemical that is harmless when cool may be dangerous when heated.

17. Maintain a clean work area and keep all materials away from flames.

18. Never reach across a flame.

19. Make sure you know how to light a Bunsen burner. (Your teacher will demonstrate the proper procedure for lighting a burner.) If the flame leaps out of a burner toward you, immediately turn off the gas. Do not touch the burner. It may be hot. And never leave a lighted burner unattended!

20. When heating a test tube or bottle, always point it away from you and others. Chemicals can splash or boil out of a heated test tube.

21. Never heat a liquid in a closed container. The expanding gases produced may blow the container apart, injuring you or others.

22. Before picking up a container that has been heated, first hold the back of your hand near it. If you can feel the heat on the back of your hand, the container may be too hot to handle. Use a clamp or tongs when handling hot containers.

Using Chemicals Safely

23. Never mix chemicals for the "fun of it." You might produce a dangerous, possibly explosive substance.

24. Never touch, taste, or smell a chemical unless you are instructed by your teacher to do so. Many chemicals are poisonous. If you are instructed to note the fumes in an experiment, gently wave your hand over the opening of a container and direct the fumes toward your nose. Do not inhale the fumes directly from the container.

25. Use only those chemicals needed in the activity. Keep all lids closed when a chemical is not being used. Notify your teacher whenever chemicals are spilled.

26. Dispose of all chemicals as instructed by your teacher. To avoid contamination, never return chemicals to their original containers.

27. Be extra careful when working with acids or bases. Pour such chemicals over the sink, not over your workbench.

28. When diluting an acid, pour the acid into water. Never pour water into an acid.

29. Immediately rinse with water any acids that get on your skin or clothing. Then notify your teacher of any acid spill.

Using Glassware Safely

30. Never force glass tubing into a rubber stopper. A turning motion and lubricant will be helpful when inserting glass tubing into rubber stoppers or rubber tubing. Your teacher will demonstrate the proper way to insert glass tubing.

31. Never heat glassware that is not thoroughly dry. Use a wire screen to protect glassware from any flame.

32. Keep in mind that hot glassware will not ap-

pear hot. Never pick up glassware without first checking to see if it is hot. See #22.

33. If you are instructed to cut glass tubing, fire-polish the ends immediately to remove sharp edges.

34. Never use broken or chipped glassware. If glassware breaks, notify your teacher and dispose of the glassware in the proper trash container.

35. Never eat or drink from laboratory glassware. Thoroughly clean glassware before putting it away.

Using Sharp Instruments

36. Handle scalpels or razor blades with extreme care. Never cut material toward you; cut away from you.

37. Immediately notify your teacher if you cut your skin when working in the laboratory.

Animal Safety

38. No experiments that will cause pain, discomfort, or harm to mammals, birds, reptiles, fishes, and amphibians should be done in the classroom or at home.

39. Animals should be handled only if necessary. If an animal is excited or frightened, pregnant, feeding, or with its young, special handling is required.

40. Your teacher will instruct you as to how to handle each animal species that may be brought into the classroom.

41. Clean your hands thoroughly after handling animals or the cage containing animals.

End-of-Experiment Rules

42. After an experiment has been completed, clean up your work area and return all equipment to its proper place.

43. Wash your hands after every experiment.

44. Turn off all burners before leaving the laboratory. Check that the gas line leading to the burner is off as well.

NAME	SYMBOL	ATOMIC NUMBER	ATOMIC MASS†	NAME	SYMBOL	ATOMIC NUMBER	ATOMIC MASS†
Actinium	Ac	89	(227)	Neodymium	Nd	60	144.2
Aluminum	Al	13	27.0	Neon	Ne	10	20.2
Americium	Am	95	(243)	Neptunium	Np	93	(237)
Antimony	Sb	51	121.8	Nickel	Ni	28	58.7
Argon	Ar	18	39.9	Niobium	Nb	41	92.9
Arsenic	As	33	74.9	Nitrogen	N	7	14.01
Astatine	At	85	(210)	Nobelium	No	102	(255)
Barium	Ba	56	137.3	Osmium	Os	76	190.2
Berkelium	Bk	97	(247)	Oxygen	O	8	16.00
Beryllium	Be	4	9.01	Palladium	Pd	46	106.4
Bismuth	Bi	83	209.0	Phosphorus	P	15	31.0
Boron	B	5	10.8	Platinum	Pt	78	195.1
Bromine	Br	35	79.9	Plutonium	Pu	94	(244)
Cadmium	Cd	48	112.4	Polonium	Po	84	(210)
Calcium	Ca	20	40.1	Potassium	K	19	39.1
Californium	Cf	98	(251)	Prascodymium	Pr	59	140.9
Carbon	C	6	12.01	Promethium	Pm	61	(145)
Cerium	Ce	58	140.1	Protactinium	Pa	91	(231)
Cesium	Cs	55	132.9	Radium	Ra	88	(226)
Chlorine	Cl	17	35.5	Radon	Rn	86	(222)
Chromium	Cr	24	52.0	Rhenium	Re	75	186.2
Cobalt	Co	27	58.9	Rhodium	Rh	45	102.9
Copper	Cu	29	63.5	Rubidium	Rb	37	85.5
Curium	Cm	96	(247)	Ruthenium	Ru	44	101.1
Dysprosium	Dy	66	162.5	Samarium	Sm	62	150.4
Einsteinium	Es	99	(254)	Scandium	Sc	21	45.0
Erbium	Er	68	167.3	Selenium	Se	34	79.0
Europium	Eu	63	152.0	Silicon	Si	14	28.1
Fermium	Fm	100	(257)	Silver	Ag	47	107.9
Fluorine	F	9	19.0	Sodium	Na	11	23.0
Francium	Fr	87	(223)	Strontium	Sr	38	87.6
Gadolinium	Gd	64	157.2	Sulfur	S	16	32.1
Gallium	Ga	31	69.7	Tantalum	Ta	73	180.9
Germanium	Ge	32	72.6	Technetium	Tc	43	(97)
Gold	Au	79	197.0	Tellurium	Te	52	127.6
Hafnium	Hf	72	178.5	Terbium	Tb	65	158.9
Helium	He	2	4.00	Thallium	Tl	81	204.4
Holmium	Ho	67	164.9	Thorium	Th	90	232.0
Hydrogen	H	1	1.008	Thulium	Tm	69	168.9
Indium	In	49	114.8	Tin	Sn	50	118.7
Iodine	I	53	126.9	Titanium	Ti	22	47.9
Iridium	Ir	77	192.2	Tungsten	W	74	183.9
Iron	Fe	26	55.8	Unnilennium	Une	109	(266?)
Krypton	Kr	36	83.8	Unnilhexium	Unh	106	(263)
Lanthanum	La	57	138.9	Unniloctium	Uno	108	(265)
Lawrencium	Lr	103	(256)	Unnilpentium	Unp	105	(262)
Lead	Pb	82	207.2	Unnilquadium	Unq	104	(261)
Lithium	Li	3	6.94	Unnilseptium	Uns	107	(262)
Lutetium	Lu	71	175.0	Uranium	U	92	238.0
Magnesium	Mg	12	24.3	Vanadium	V	23	50.9
Manganese	Mn	25	54.9	Xenon	Xe	54	131.3
Mendelevium	Md	101	(258)	Ytterbium	Yb	70	173.0
Mercury	Hg	80	200.6	Yttrium	Y	39	88.9
Molybdenum	Mo	42	95.9	Zinc	Zn	30	65.4
				Zirconium	Zr	40	91.2

†Numbers in parentheses give the mass number of the most stable isotope.

Appendix E

Key

6	Atomic number
C	Element's symbol
Carbon	Element's name
12.011	Atomic mass

Transition Metals

1	**2**	**3**	**4**	**5**	**6**	**7**	**8**	**9**
1 **H** Hydrogen 1.00794								
3 **Li** Lithium 6.941	4 **Be** Beryllium 9.0122							
11 **Na** Sodium 22.990	12 **Mg** Magnesium 24.305							
19 **K** Potassium 39.098	20 **Ca** Calcium 40.08	21 **Sc** Scandium 44.956	22 **Ti** Titanium 47.88	23 **V** Vanadium 50.94	24 **Cr** Chromium 51.996	25 **Mn** Manganese 54.938	26 **Fe** Iron 55.847	27 **Co** Cobalt 58.9332
37 **Rb** Rubidium 85.468	38 **Sr** Strontium 87.62	39 **Y** Yttrium 88.9059	40 **Zr** Zirconium 91.224	41 **Nb** Niobium 92.91	42 **Mo** Molybdenum 95.94	43 **Tc** Technetium (98)	44 **Ru** Ruthenium 101.07	45 **Rh** Rhodium 102.906
55 **Cs** Cesium 132.91	56 **Ba** Barium 137.33	57 to 71	72 **Hf** Hafnium 178.49	73 **Ta** Tantalum 180.95	74 **W** Tungsten 183.85	75 **Re** Rhenium 186.207	76 **Os** Osmium 190.2	77 **Ir** Iridium 192.22
87 **Fr** Francium (223)	88 **Ra** Radium 226.025	89 to 103	104 **Unq** Unnilquadium (261)	105 **Unp** Unnilpentium (262)	106 **Unh** Unnilhexium (263)	107 **Uns** Unnilseptium (262)	108 **Uno** Unniloctium (265)	109 **Une** Unnilennium (266)

Rare-Earth Elements

Lanthanoid Series

57 **La** Lanthanum 138.906	58 **Ce** Cerium 140.12	59 **Pr** Praseodymium 140.908	60 **Nd** Neodymium 144.24	61 **Pm** Promethium (145)	62 **Sm** Samarium 150.36

Actinoid Series

89 **Ac** Actinium 227.028	90 **Th** Thorium 232.038	91 **Pa** Protactinium 231.036	92 **U** Uranium 238.029	93 **Np** Neptunium 237.048	94 **Pu** Plutonium (244)

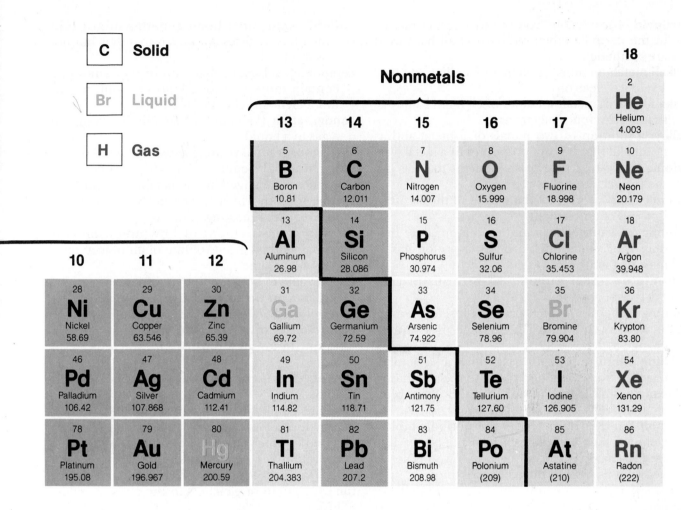

	Solid
C	

	Liquid
Br	

	Gas
H	

Nonmetals

18

2
He
Helium
4.003

13	**14**	**15**	**16**	**17**
5	6	7	8	9
B	**C**	**N**	**O**	**F**
Boron	Carbon	Nitrogen	Oxygen	Fluorine
10.81	12.011	14.007	15.999	18.998

10
Ne
Neon
20.179

10	**11**	**12**	13	14	15	16	17	18
			Al	**Si**	**P**	**S**	**Cl**	**Ar**
			Aluminum	Silicon	Phosphorus	Sulfur	Chlorine	Argon
			26.98	28.086	30.974	32.06	35.453	39.948
28	29	30	31	32	33	34	35	36
Ni	**Cu**	**Zn**	**Ga**	**Ge**	**As**	**Se**	**Br**	**Kr**
Nickel	Copper	Zinc	Gallium	Germanium	Arsenic	Selenium	Bromine	Krypton
58.69	63.546	65.39	69.72	72.59	74.922	78.96	79.904	83.80
46	47	48	49	50	51	52	53	54
Pd	**Ag**	**Cd**	**In**	**Sn**	**Sb**	**Te**	**I**	**Xe**
Palladium	Silver	Cadmium	Indium	Tin	Antimony	Tellurium	Iodine	Xenon
106.42	107.868	112.41	114.82	118.71	121.75	127.60	126.905	131.29
78	79	80	81	82	83	84	85	86
Pt	**Au**	**Hg**	**Tl**	**Pb**	**Bi**	**Po**	**At**	**Rn**
Platinum	Gold	Mercury	Thallium	Lead	Bismuth	Polonium	Astatine	Radon
195.08	196.967	200.59	204.383	207.2	208.98	(209)	(210)	(222)

The symbols shown here for elements 104-109 are being used temporarily until names for these elements can be agreed upon.

Metals

Mass numbers in parentheses are those of the most stable or common isotope.

63	64	65	66	67	68	69	70	71
Eu	**Gd**	**Tb**	**Dy**	**Ho**	**Er**	**Tm**	**Yb**	**Lu**
Europium	Gadolinium	Terbium	Dysprosium	Holmium	Erbium	Thulium	Ytterbium	Lutetium
151.96	157.25	158.925	162.50	164.93	167.26	168.934	173.04	174.967
95	96	97	98	99	100	101	102	103
Am	**Cm**	**Bk**	**Cf**	**Es**	**Fm**	**Md**	**No**	**Lr**
Americium	Curium	Berkelium	Californium	Einsteinium	Fermium	Mendelevium	Nobelium	Lawrencium
(243)	(247)	(247)	(251)	(254)	(257)	(258)	(259)	(260)

Glossary

actinoid series: second row of rare-earth elements in the periodic table; radioactive; all but three are synthetic

alkali metal: member of element Family 1 that has 1 valence electron

alkaline earth metal: member of element Family 2 that has 2 valence electrons

alloy: a solution of two metals or a metal and a nonmetal that has the properties of a metal

atom: smallest particle of an element that has all the properties of that element

atomic mass: average of the masses of the existing isotopes of an element

atomic mass unit (amu): unit used to measure the masses of subatomic particles; a proton has a mass of 1 amu

atomic number: number of protons in the nucleus of an atom

boiling: process in which particles inside a liquid as well as those on the surface of a liquid change to a gas

boiling point: temperature at which a substance changes from the liquid phase to the gas phase

boron family: Family 13 of the periodic table; elements have 3 valence electrons

carbon family: Family 14 of the periodic table; elements have 4 valence electrons

chemical change: process by which a substance becomes a new and different substance

chemical equation: expression in which symbols, formulas, and numbers are used to represent a chemical reaction

chemical formula: combination of chemical symbols usually used to represent a compound

chemical property: property that describes how a substance changes into a new substance

chemical reaction: process in which the physical and chemical properties of the original substance change as a new substance with different physical and chemical properties is formed

chemical symbol: shorthand way of representing an element

coefficient (koh-uh-FIHSH-uhnt): number that is placed in front of a symbol or a formula in a chemical equation that indicates how many atoms or molecules of this substance are involved in the reaction

colloid (KAHL-oid): homogeneous mixture in which the particles are mixed together but not dissolved

compound: substance made up of molecules that contain more than one kind of atom; two or more elements chemically combined

condensation (kahn-duhn-SAY-shuhn): change of a gas to a liquid

corrosion: gradual wearing away of a metal due to a chemical reaction in which the metal element is changed into a metallic compound

crystal: solid in which the particles are arranged in a regular, repeating pattern

density: measurement of how much mass is contained in a given volume of an object; mass per unit volume

ductile: able to be drawn into a thin wire

electromagnetic force: force of attraction or repulsion between particles in an atom

electron: negatively charged subatomic particle found in an area outside the nucleus of an atom

electron cloud: space in which electrons are likely to be found

element: simplest type of pure substance

energy level: most likely location in an electron cloud in which an electron can be found

evaporation (ee-vap-uh-RAY-shuhn): vaporization that takes place at the surface of a liquid

family: column of elements in the periodic table; group

flammability (flam-uh-BIHL-uh-tee): ability to burn

freezing: change of a liquid into a solid

freezing point: temperature at which a substance changes from the liquid phase to the solid phase

gas: phase in which matter has no definite shape or volume

gravity: force of attraction between all objects in the universe

group: column of elements in the periodic table; family

halogen family: Family 17 of the periodic table; elements have atoms that contain 7 valence electrons

heterogeneous (heht-er-oh-JEE-nee-uhs) **mixture:** substance that does not appear to be the same throughout

homogeneous (hoh-moh-JEE-nee-uhs) **mixture:** mixture that appears the same throughout

inertia (ihn-ER-shuh): tendency of objects to remain in motion or to stay at rest unless acted upon by an outside force

insoluble: unable to be dissolved in water

isotope (IGH-suh-tohp): atom of an element that has the same number of protons as another atom of the same element but a different number of neutrons

lanthanoid series: first row of rare-earth elements in the periodic table; soft, malleable metals that have a high luster and conductivity

liquid: matter with no definite shape but with a definite volume

luster: shininess

malleable: able to be hammered out into a thin sheet

mass: amount of matter in an object

mass number: sum of the protons and neutrons in the nucleus of an atom

matter: anything that has mass and volume

melting: change of a solid to a liquid

melting point: temperature at which a substance changes from the solid phase to the liquid phase

metal: element that is a good conductor of heat and electricity, is shiny, has a high melting point, is ductile and malleable, and tends to lose electrons

metalloid (MEHT-uh-loid): element that has properties of both metals and nonmetals

mixture: matter that consists of two of more substances mixed but not chemically combined

molecule (MAHL-ih-kyool): structure made up of two or more atoms

neutron: subatomic particle with no electric charge that is found in the nucleus of an atom

nitrogen family: Family 15 of the periodic table; elements have atoms with 5 valence electrons

noble gas: member of Family 18 of the periodic table; elements have atoms with 8 valence electrons and are extremely unreactive

nonmetal: element that is a poor conductor of heat and electricity, has a dull surface, low melting point, is brittle, breaks easily, and tends to gain electrons

nucleus (NOO-klee-uhs): small, dense positively charged center of an atom

oxygen family: Family 14 of the periodic table; elements have atoms with 6 valence electrons

period: horizontal row of elements in the periodic table

periodic law: law that states that the physical and chemical properties of the elements are periodic functions of their atomic numbers

phase: state in which matter can exist: solid, liquid, gas, or plasma

physical property: characteristic that distinguishes one type of matter from another and can be observed without changing the identity of the substance

plasma: phase in which matter is extremely high in energy and cannot be contained by ordinary matter; very rare on Earth

property: characteristic of a substance

proton: subatomic particle that has a positive charge and is found in the nucleus of an atom

pure substance: substance made of one kind of material having definite properties

quark (KWORK): particle that makes up all other known particles in the nucleus of an atom

rare-earth element: general designation for those elements in the lanthanoid and actinoid series

solid: phase in which matter has a definite shape and volume

solubility: the amount of a solute that can be completely dissolved in a given amount of solvent at a specific temperature

soluble (SAHL-yoo-buhl): can be dissolved in water

solute (SAHL-yoot): substance that is dissolved in a solution

solution (suh-LOO-shuhn): homogeneous mixture in which one substance is dissolved in another

solvent (SAHL-vuhnt): substance that does the dissolving in a solution

strong force: force that binds protons and neutrons in the nucleus

subatomic particle: proton, neutron, and electron

subscript: number placed to the lower right of a chemical symbol to indicate the number of atoms of the element in the compound

sublimation (suhb-luh-MAY-shuhn): change from the solid phase directly into the gas phase

transition metal: element that has properties similar to other transition metals and to other metals but whose properties do not fit in with those of any other family

vaporization (vay-per-ih-ZAY-shuhn): change of a liquid to a gas

volume: amount of space an object takes up

weak force: force that is the key to the power of the sun; responsible for a process known as radioactive decay

weight: measure of the force of attraction between objects due to gravity

Matter
    atoms, N85–101
    classification of, N58–59
    compounds, N71–76
    definition of, N21
    density, N21–25
    elements, N68–71
    mass, N14–16
    mixtures, N60–64
    phases of, N34–47
    properties of, N12–13
    solutions, N64–68
    volume, N21
    weight, N17–20
Melting point, N42–43
Mendeleev, Dmitri, N108–111
Mercury pollution of waters,
    N135–136
Metalloids, N121, N125
    boron family, N125
    carbon family, N125–126
    properties of, N121
Metals, N118–120, N122–124
    alkali metals, N122–123
    alkaline earth metals, N123
    boron family, N125–126
    carbon family, N125–126
    chemical properties of,
        N119–120
    physical properties of, N118
    transition metals, N124–125
Milliliter (mL), N21
Mixtures, N60–64
    colloids, N63–64
    definition of, N60
    heterogeneous mixtures, N63
    homogeneous mixtures, N63–64
    properties of, N60–62
Models
    computer generated, N91
    nature of, N84
Molecules, of compounds, N73
Moseley, Henry, N111–112

Neutrons, N93
Newton (N), N20
Nitrogen family, N127
Noble gases, N129
Nonmetals, N120–121
    carbon family, N125–126
    chemical properties of,
        N120–121
    halogen family, N128–129
    nitrogen family, N127
    oxygen family, N127–128
    physical properties of, N120
Nucleus of atom, N88–89, N92–93
    neutrons, N93
    protons, N92–93
    size of, N92

Oxygen family, N127–128

Period, of periodic table, N116
Periodic law, N111
Periodic table, N110-130
    atomic numbers, discovery of,
        N111
    boron family, N125-126
    carbon family, N126–127
    columns of elements in, N113,
        N116
    groups or families of elements,
        N113, N116
    halogen family, N128–129
    key to information about
    elements, N117–118
    Mendeleev's arrangement of,
        N108–110
    metalloids, N121, N125
    metals, N118–120, N122–124
    modern periodic table,
        N111–112
    nitrogen family, N127
    noble gases, N129
    nonmetals, N120–121
    oxygen family, N127–128
    periodic properties of elements,
        N131–133
    periods in, N116
    rare-earth elements, N130
    rows in, N116
    transition metals, N124–125
Phase, meaning of, N34
Phase changes
    boiling, N44–45
    condensation, N45
    and energy, N41–42
    evaporation, N44
    freezing, N43
    liquid-gas phase changes,
        N44–45
    melting, N42–43
    solid-gas phase changes, N46–47
    solid-liquid phase changes,
        N42–43
    sublimation, N46–47
    vaporization, N44
Phases of matter, N34–47
    gases, N37–40
    liquids, N36
    plasma, N40–41
    solids, N34–36
Plasma, N40–41
    characteristics of, N40–41
Plum pudding model of the atom,
    N87–88
Properties
    chemical properties, N48–50
    of matter, N12–13
    meaning of, N12
    physical properties, N34

Protons, N92–93
    and atomic number, N93–94
Pure substances, N68–69
    compounds as, N72
    elements as, N69
    nature of, N68–69

Quarks, N98

Rare-earth elements, N130
    actinoid series, N130
    lanthanoid series, N130
Rutherford's model of the atom,
    N88–89

Solids, N34–36
    amorphous solids, N34–35
    characteristics of, N34
    crystalline solids, N34
    solid-gas phase changes, N46–47
    solid-liquid phase changes,
        N42–43
Solutions, N64–68
    alloys, N66–67
    definition of, N64
    solubility, N66
    solute of, N64
    solvent of, N64
Strong force, N100
Subatomic particles, N92
    atomic forces of, N99–100
    electrons, N96–98
    measurement of, N93
    neutrons, N93
    protons, N92–93
Sublimation, N46–47
Subscripts, in chemical formulas,
    N74

Thomson's model, of the atom,
    N87–88
Transition metals, N124–125

Valence number, of element, N132
Vaporization, N44
Viscosity, N36
Volume, N21
    definition of, N21
    of liquids, N36
    measurement of, N21

Waste disposal, N51
Wave mechanics, theory of,
    N89–90
Wave model of the atom, N89–90
Weak force, N100
Weight, N17–20
    changeability of, N17-18
    and gravity, N18-20
    measurement of, N20